George Cunningham Edwards

Elements of geometry

George Cunningham Edwards

Elements of geometry

ISBN/EAN: 9783337276935

Printed in Europe, USA, Canada, Australia, Japan

Cover: Foto ©berggeist007 / pixelio.de

More available books at **www.hansebooks.com**

ELEMENTS OF GEOMETRY

BY

GEORGE C EDWARDS, Ph.B.

Associate Professor of Mathematics in the University of
California

New York
MACMILLAN AND CO.
AND LONDON
1895

PREFACE.

THE large number of Geometries recently published is an indication of the fact that, so far as Geometry is concerned, there is an unsatisfied want in High Schools, Academies, and Colleges. It is in the hope of supplying this need that the book here presented has been written.

Effort has been made so to frame the definitions that they will not have to be changed when the student comes to the higher reaches of the subject, or when he advances to other branches of mathematics. Effort has been made to introduce new material when needed, and not before; to appeal to the understanding; to discourage mere memorizing; and to avoid entering into confusing details.

Corollaries and scholia have been in large part replaced by exercises, every one of which the student must work out as he comes to them.

At the end of the Plane Geometry and at the end of the Solid Geometry there will be found a sufficiently large number of exercises to give a review of the work preceding them, and thoroughly to establish method of attack in the mind of the student. The exercises which have been taken from other authors are limited to such as appear in at least two of them.

Simplicity in wording and in demonstration have been sought; numerous notes are added; leading thoughts, definitions, and methods are repeated so as to enforce them; the student is led, not driven; and he is encouraged to investigate and determine for himself.

The division of the work into fourteen chapters makes a natural arrangement of parts, although the chapters are quite unequal in length. The chapters have been divided into articles, and these have been numbered for convenience in cross-reference. But the student should refer to principles or facts by stating them in full.

The writer thinks that the proper place for the theory of proportion is in the Algebra, and for that reason has omitted it here.

The leading features of the book are, the development of, and the insistence on, *method of attack* in the solution of problems.

Logic, which is the science of orderly thinking, requires for its proper study, a development of mind beyond that which is necessary for the student of the ordinary Geometry. Geometry, however, does admirably illustrate the principles of Logic; and in it these principles are rigidly enforced at every step without the distractions that accompany investigations in any other field of research. From the educational point of view Geometry is *the most valuable* study of the High School period.

Instructors are advised to proceed slowly with the earlier chapters, and more rapidly with the later ones; not abating at all in thoroughness.

Instructors and students are advised to read carefully the paragraphs on Demonstrative Geometry, pages 112–116 of the "Report of the Committee on Secondary School Studies, appointed at the meeting of the National Educational Association, July 9, 1892"; and published by the United States Bureau of Education in 1893.

The writer will deem it a favor to have his attention called to errors. They will surely exist in the first writing, but the author hopes that the main features of the book will not be seriously injured thereby.

GEORGE C. EDWARDS.

BERKELEY, March 23, 1895.

CONTENTS.

	PAGE
INTRODUCTION	xv

CHAPTER I.

Definitions	1
Direction axiom	2
Rotation axiom	4
An angle	5
Figures	6
Translation axiom	6
Equality axioms	12
Perpendiculars	14

CHAPTER II.

A triangle	17
Parallels	21
Perpendicular bisector	24
A circle	25
Congruent triangles	26
Construction of triangles	28
Inequality axioms	32

CHAPTER III.

Triangles	35
Quadrilaterals and quadrangles	42

	PAGE
Polygons	46
Analysis	48

CHAPTER IV.

A circle	51
Angles and arcs	54
Triangles and circles	64

CHAPTER V.

Measurement of distance	69
Area	71
Proportional division	76
Medians	83

CHAPTER VI.

Squares on segments	87
Squares on the sides of triangles	89
Areas of similar triangles	95
Areas of similar polygons	96
Metrical relations of parts of a triangle	99

CHAPTER VII.

Chords and tangents	103

CHAPTER VIII.

Inscribed and circumscribed polygons	117
Variable and limit	121
Limit axioms	124
Area of a circle	126
PROBLEMS	137

CHAPTER IX.

Intersections of planes	175
Perpendiculars to planes	177
Parallels to planes	184

CHAPTER X.

A sphere	188
Plane sections of a sphere	188
Spherical arcs	189
Spherical triangles	192

CHAPTER XI.

Triedrals	205

CHAPTER XII.

Prisms	210
Pyramids	210
Cylinders	213
Cones	219
Surface of a sphere	227
Zone	229
Lune	230
Area of a spherical triangle	231
Areas of similar surfaces	232

CHAPTER XIII.

Volume of a prism	236
Volume of a pyramid	241
Prismoidal formula	245
Truncated prism	246
Volume of a sphere	249
Volumes of similar figures	251

CONTENTS.

CHAPTER XIV.

	PAGE
The parabola	254
Line relations	254
Area of a parabola	262
Particular cases	264
The ellipse	265
Line relations	266
Area of an ellipse	270
Particular cases	272
The hyperbola	272
Line relations	274
Particular cases	281
PROBLEMS	283

SYMBOLS.

$\angle$,	Angle.
$\angle\!\!\!\angle$,	Angles.
$\triangle$,	Plane triangle.
$\triangle\!\!\!\triangle$,	Plane triangles.
$=$,	Equals, or is equal to.
$\parallel$,	Parallel.
$<$,	Less than.
$>$,	Greater than.
$\square$,	Parallelogram.
$\square\!\!\!\square$,	Parallelograms.
$\perp$,	Perpendicular.
$\odot$,	Circle, or circumference (depending on the context).
sp. $\triangle$,	Spherical triangle.
$\overline{PQ}$,	The straight line segment PQ.
$\overset{\frown}{PQ}$,	The arc PQ.
Q. E. D.,	Quod Erat Demonstrandum (which was to be proved).
Q. E. F.,	Quod Erat Faciendum (which was to be done).
$\therefore$,	Hence.

INTRODUCTION.

GEOMETRY had its origin, as the name indicates, in the need of a method for the accurate description of limited portions of the Earth's surface.

At the present time the development of the science is such as to include a vast amount of material which is beyond the scope of a book devoted to the consideration of elementary forms in such space as that in which we live and exercise our senses.

Like any other science, *Geometry* is built upon definitions and axioms. *Definitions* describe, in as simple a manner as possible, the objects with which we have to deal. *Axioms* are self-evident truths that relate to the objects described, and the operations to be performed. An *axiom* is a truth self-evident to one who understands the terms in which it is stated.

Upon the definitions and axioms of Geometry is built up our knowledge of fundamental relations. Upon these more complex relations are based. Upon these new relations, still more advanced relations are built; and thus the science is developed. As is true of every other science, the extent to which Geometry may be pursued is without limit.

While the beginnings are based upon definitions and axioms, it does *not* follow that *all* of the definitions and axioms are to be brought forward at the opening of the subject. As the information of the student grows, new definitions and new axioms are appropriate.

Advance in Geometry is made through an orderly arrangement of theorems and problems.

A *theorem* is a general statement of relations. These relations are established by a course of reasoning.

A *problem* is a demand that a construction be made; that certain relations be established; or that relations between certain things be established.

ELEMENTS OF GEOMETRY.

CHAPTER I.

1. Definitions. An enclosed portion of space, no matter what its form or material, or if it exist in the imagination only, is called a **volume**.

That which separates a volume from excluded space is called **surface**, and does not partake of the character of the material, if there be any used. The surface that

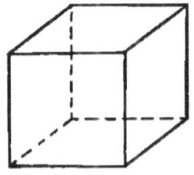 

FIG. 1.

separates the hull of a ship from the water is neither water nor the material of which the ship may be constructed.

2. Definitions. A position in space that is without magnitude is called a **point**.

If a *point* move it will generate a line; it may move at a snail's pace, or it may move with the rapidity of thought.

2 ELEMENTS OF GEOMETRY.

The position of one point with respect to another determines *direction*. If we represent the position of one point by the letter (*A*), placed near it, and the position of another point by the letter (*B*), placed near it, the direction (*AB*) is established. The direction of (*B*) from (*A*) and the direction of (*A*) from (*B*) are opposites.

Relations expressed by the words *same* and *opposite*, whether they be of interpretation or of operation, are expressed symbolically by the vertical cross (+), and by the horizontal dash (−). If (+4) represents a number of miles in the direction of (*B*) from (*A*), (−4) will represent the same distance in the direction of (*A*) from (*B*). If (+6) represents six years to come, (−4) will represent four years that have passed.

Anything which is so large that it is beyond our comprehension, is said to be *infinitely* large. Space is *infinitely* large; the number of points in space is *infinite*; time is *infinite*.

DIRECTION AXIOM. *From any point in space there will be an infinite number of directions; and each direction will have its opposite direction.*

There will also be an infinite number of points in any assumed direction from a given point, which points will each be at different distances from the given point. In each opposite direction there will also be an infinite number of points. The rays of light from

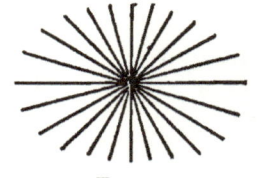

FIG. 2.

an arc-light or from a fixed star are fairly good illustrations.

DEFINITIONS. 3

3. If a point move from an assumed position, it may start in any one of an infinite number of directions. If it continue in the direction in which it starts, so as to pass through the position of every point lying in the same direction (§ 2), it will generate what is called a *straight line*.

Since a straight line is determined by direction, and two points determine direction, two points determine the position of a straight line.

If *separate* straight lines have any point in common there can be but *one*, for if they have *two* coincident points the two lines will form *one* line.

The straight line is infinite in extent.

The portion of a straight line between any two points of the line is called a **segment**.

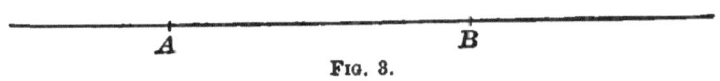

Fig. 3.

Unless otherwise specified, the indefinite straight line is to be understood when two points are named which locate the line, as: the line AB (the segment is represented by $\overline{AB}$).

An infinite number of straight lines may pass through a point.

If a point move with ever-changing direction, it will generate a *curved line*; the law of change determining the character of the *curve*.

As any fixed point has a fixed direction from another fixed point, only one straight line can join two points.

4. Our idea of distance involves the ideas of motion and of time. Time *is*, and advances with uniformity. Motion may or may not be uniform. If motion be uniform, the greater the time the greater the distance. In equal times the more rapid of two motions will cover the greater distance.

If a point move from position A to position B without change of direction, it will reach position B after having moved over a less distance, than if on the way it had made any changes of direction.

We therefore say that the shortest distance between two points is that portion of the straight line determined by them (§ 3) which lies between them.

Although the straight lines with which we deal are in general infinite in length, our representations of them are limited by the extent of the surface upon which our representations are made.

Henceforth when the word *line* is used without any qualification a *straight* line is understood.

5. ROTATION AXIOM. *If two straight lines intersect, either or both may be moved without changing the point of intersection. They may be brought to coincide; in which case all the points of the two lines are in common. The lines are then said to be congruent.*

PLANE GEOMETRY.

6. A **plane** is a surface such that if *any* two of its points are joined by a straight line it will contain the entire line.

An infinite number of straight lines may lie in a plane.

In drawing, a dot upon a plane is used as the representative of a point, and a mark by chalk, pencil, or pen is used to represent a line.

AN ANGLE.

7. Let the surface of this page represent a plane, and AB a line of the plane.

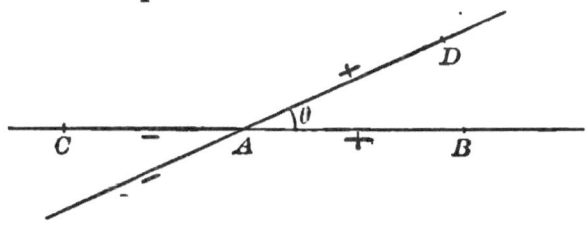

Fig. 4.

If the direction of B from A be positive ($+$), the direction of C from A will be negative ($-$).

If the straight line rotate about the point A as a pivot, and remain in the plane of the page, it will make a change in direction. If the motion be considered as having stopped when the line has arrived at the position AD, the change in direction is called an **angle**; and A is called its **vertex**.

6   ELEMENTS OF GEOMETRY.

The angle may be described as: the angle *BAD* (∠ *BAD*), or by a single symbol as θ.

The distances *AB* or *AD* have nothing whatever to do with the magnitude of the angle; it is purely a matter of direction.

FIGURES.

8. Volumes, surfaces, lines, angles, and points, or any combination of them that we may make, constitute **geometric figures**.

The relations of parts of the same figure to each other, and the relations of different figures to each other, constitute the subject matter of geometry.

The Elementary Geometry is ordinarily separated into two parts, viz.:

(*a*) That which relates to figures in a plane, and is called **Plane Geometry**.

(*b*) That which relates to figures that do not lie in one plane only, and is called **Solid Geometry**.

From now on until we arrive at § 105, we shall be concerned with Plane Geometry.

TRANSLATION AXIOM. *A geometric figure may be moved at pleasure without changing the relations existing between the parts which compose it.*

MORE ABOUT PLANES.

9. A plane may be made to pass through a given point: for a plane may be moved so as to cause any one of its points to coincide with the given point.

Any infinite number of planes may be made to pass through a given point: because the manner of moving any plane so as to cause it to contain a given point has not been limited.

If a plane pass through a point, any line which contains the given point, and lies in the plane, may, by motion of the plane, be brought to pass through a second point, located anywhere.

Thus we see that a plane may be passed through any two points. It will contain the straight line joining those points. Every possible position of a plane which contains two given points, may be reached by causing a selected plane which contains them to rotate about the line joining them as an axis, until it returns to the position it had at starting. Thus we see that an infinite number of planes may be passed through two points. These planes will each contain the line determined by the two points.

A plane passed through a line and making a complete rotation on the line as an axis, will encounter every point in space.

10. If a line in each of two planes be made to coincide, and then one of the planes be rotated about this line until at least *one* point, not in the common line, shall be common to the two planes, any line which may be drawn through this point and any point of the common line will lie in both planes. Lines may therefore be drawn through this point so as to reach every point of both planes; in which case the two planes would have *every* point in common. Hence the statement or

8   ELEMENTS OF GEOMETRY.

THEOREM. *Two planes may be made to coincide.*

Remark. — By virtue of this possibility all planes are said to be congruent.

11. If the line AB remain in the plane of the page, and be revolved about A as a pivot, it will generate an angle by its change of direction (§ 7). When the positive direction AB has changed to that of AC, the angle BAC ($\angle BAC$) will have been generated. When it shall have been still further changed to the direction AD, $\angle BAD$ will be the result. The rotation may still continue, and the positive direction may become in succes-

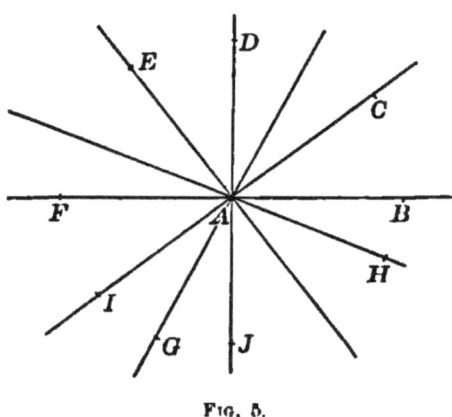

FIG. 5.

sion, AE, AF, AG, AH, and may finally coincide with its first position, AB.

If the direction AB be positive, the direction AF is negative. When the positive direction has been changed to AC, the negative has been changed to AI. When the positive has been changed to AD, the negative has been changed to *AJ*. When the positive has arrived at *AF*, the negative has arrived at *AB*. And when the positive has again come to the position *AB*, the negative will have returned to its corresponding position, *AF*.

Both the positive and the negative directions have generated angles, and each has made a complete rotation. In this rotation *each* has passed through *every* point in the plane. Since all positions are thus merely duplicated, the angle generated by the positive direction is, in general, the only one considered.

For convenience of description and measurement the complete rotation is separated into parts; generally into 360 equal parts, each being called a **degree**; sometimes into 400 equal parts, each being called a **grade**.

Degrees are again separated into 60 equal parts, each called a **minute**; each minute is separated into 60 equal parts, called **seconds**; and each second is then separated decimally, if a further division is necessary.

Grades are separated decimally.

Each system of division and subdivision has its advantages. The sexagesimal is the system in common use, and has been since a period of time long antedating the Christian era. Nearly all of the literature and instruments of observation involving the consideration of angles employ this system.

But the decimal system accords with our method of enumeration, and will at some time supplant the other. May the time soon come.

If the portion of the plane below the line *FB* be revolved on that line as an axis, it may be brought to coincide with that which is above the same line (§ 10); and in that position, if the line *AB* be rotated about *A* as a pivot until it reach the direction *AF*, it will have generated the same angle in the two coincident figures. Hence the two angles thus generated must be equal. Each is one-half of the angle generated by a complete rotation; and may be expressed as 180 degrees (180°), or 200 grades (200ᵍ).

The dotted line in the accompanying figure is used to indicate the rotation by marking the path of the point *B* during the generation of the angle of 360°, which is separated into two equal parts by the line *FB*.

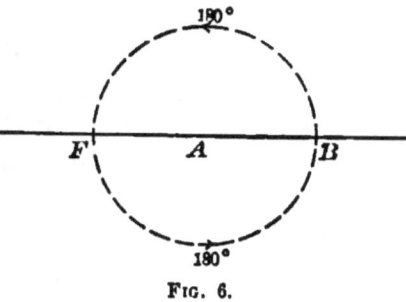

Fig. 6.

If the line *AB* change its direction 90°, it will occupy the position indicated by *AD*, and the angles *BAD* and *DAF* will be equal.

If the part of the figure to the right of the line *AD* be revolved on *AD* as an axis until the revolved portion coincides with the part on the left of *AD*, the ∠ *DAB* will coincide with the ∠ *DAF*; *AB* will fall in the direction *AF*; the ∠ *BAJ* will coincide with ∠ *FAJ*; and so must be equal to it.

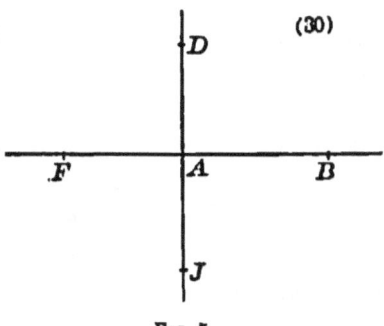

Fig. 7.

Hence (∴) the 180° of change of direction from *AF* to *AB* will be bisected by *AJ*. Thus we see that if two lines intersect so as to form *an* angle of 90°, there will be formed *four* angles of 90° each.

An angle of 90° is sometimes called a **quadrant,** but more frequently a **right angle.**

Lines at right angles with each other are said to be **perpendicular** (⊥); and either line with respect to the other is called a **perpendicular.**

When the sum of two angles is 90°, each is said to be the **complement** of the other.

When the sum of two angles is 180°, each is said to be the **supplement** of the other.

An angle that is less than 90° is said to be **acute.**

An angle greater than 90°, and less than 180°, is said to be **obtuse.**

Acute and obtuse angles are frequently spoken of as **oblique.**

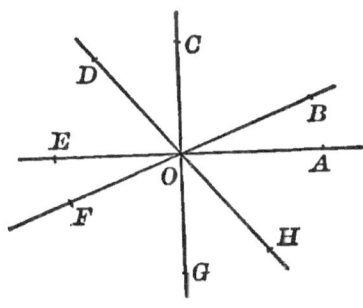

Fig. 8.

Angles which occupy toward each other the position that ∠ *AOB* and *EOF* do, are said to be **vertical.** ∠ *AOF* and *BOE* are vertical.

Note. — Although straight lines and planes both extend to infinity, it is only the relations between parts at a finite distance that concern us in the present work.

12. The sign of equality is (=).

In considering the aggregation of quantities, addition is indicated by (+), and subtraction by (−), (§ 2).

EQUALITY AXIOMS. (a) *If two things are equal to a third thing, they are equal to each other.*

(b) *The whole is greater than any of its parts, and equals the sum of all its parts.*

(c) *If the same operation be performed upon equals, the results will be equal.*

THEOREM. *If two lines intersect, the vertical angles are equal.*

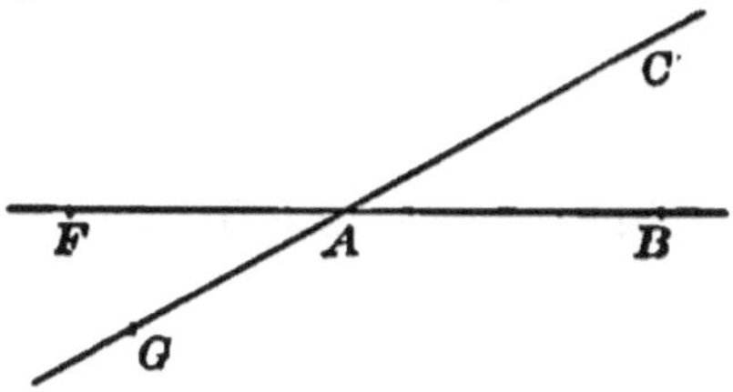

FIG. 9.

Proof. $\angle BAC + \angle CAF = 180°$, § 11.

$\angle CAF + \angle FAG = 180°$. § 11.

By Equality Axiom (a),

$\angle BAC + \angle CAF = \angle CAF + \angle FAG$.

Then subtracting $\angle CAF$ from each member of the equation, acting under the authority given by Equality Axiom (c), we have: $\angle BAC = \angle FAG$. Q. E. D.

Again, $\angle BAC + \angle CAF = 180°$,

$\angle BAC + \angle BAG = 180°$.

By Equality Axiom (a),

$\angle BAC + \angle CAF = \angle BAC + \angle BAG$.

Then subtracting $\angle BAC$ from each member of the equation, we have: $\angle CAF = \angle BAG$. Q. E. D.

ANGLES.

Exercise. — Establish the fact that $\angle BAC = \angle FAG$, by rotating one of them until it coincides with the other.

Proof. — Under the authority given by the axiom in § 8, the $\angle BAC$ may (without changing the relations between its parts) be rotated about A as a pivot. When the line AB shall have rotated 180°, the line AC will also have rotated 180°. AB will then coincide with AF, and AC with AG.

Therefore all the parts coincide and the $\angle FAG$ equals the $\angle BAC$. Q.E.D.

Note. — A mechanical device for determining approximately the value of an angle, and for the purpose of moving a given angle to any desired position, is called a protractor. A convenient form for paper or blackboard work may be made from pasteboard.

Some of the forms used are:

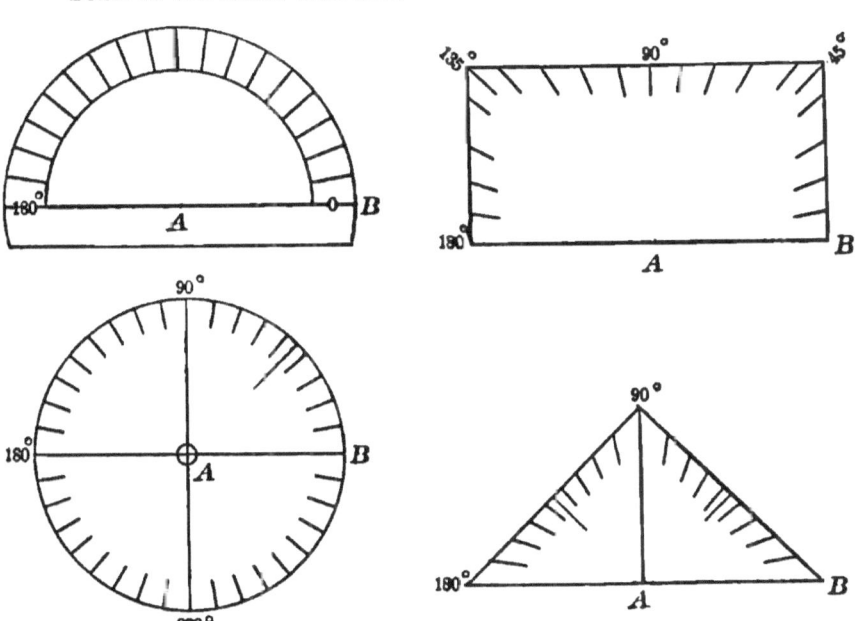

Fig. 10.

13. Theorem. *At a given point in a line one, and only one, perpendicular can be erected.*

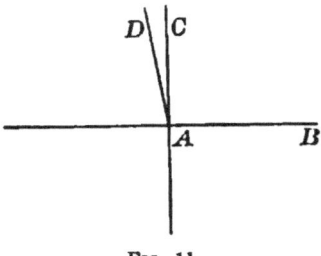

Fig. 11.

(*a*) We may at any point (*A*) of a line have at least one perpendicular: because any point of a line might serve as a pivot about which rotation could take place, and when the rotation has extended to one-fourth of a complete rotation it would be a perpendicular (§ 11).

(*b*) If two perpendiculars, as *AC* and *AD*, *could* be erected at the same point, we should have two unequal angles, each equalling 90°, which by our axiom cannot be.

Hence the Theorem is established.

Note.—The T square is a mechanical device, used in drawing, for the purpose of erecting perpendiculars at any points of a line; and, as we shall see in the next article, for letting fall perpendiculars to a line from points not in the line.

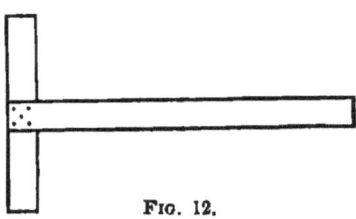

Fig. 12.

14. Theorem. *From a point not on a given straight line one, and only one, perpendicular may be drawn to the line.*

(*a*) Since at *any* point of a line a perpendicular can be erected to the line (§ 13), we may conceive of a perpendicular as moving along the line so as to be always perpendicular to it.

The moving perpendicular may be caused to pass through *any* point in the plane, as *P*; when it does so pass, there will be at least *one* perpendicular to the line *AB* through the point *P*.

(*b*) Let *BC* represent the given line, and *FA*, a perpendicular through *A*.

If any other straight line could be drawn through *A* that should be perpendicular to *BC*, let *AE* represent it.

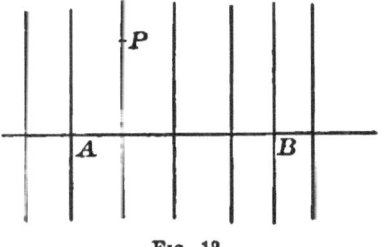

Fig. 13.

On the ⊥ *FA*, lay off *FD* = *AF*, and draw *DE*.
AE and *DE* will intersect and have but one point in common by the direction axiom.

If *AE* is a perpendicular, *DE* will also be one: because the figure below the line *BC* may be revolved on *BC* as an axis, and be made to coincide with that portion of the figure above the line; the angles remaining the same during the revolution.

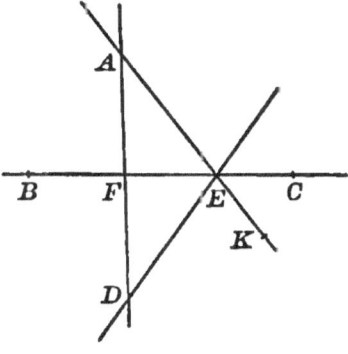

Then if *AEB* is a right angle, *DEB* will be one. But by § 11, if *AEB* is a right angle, *KEB* will also be a right angle.

And *KEB* would equal

Fig. 14.

DEB, which is in conflict with axiom (*b*) of § 12. ✔ ?

Hence the theorem is established.

NOTE.—The supposition that there could be a perpendicular through *A*, other than *AF*, to the line *BC* leads to conclusions which are in conflict with our axioms. Hence the supposition has been shown to be an erroneous one.

The student will carefully note the method employed in §§ 13 and 14. It is shown that a perpendicular through A *may* exist, and that another line through A, which shall be perpendicular to BC, cannot be drawn.

This method is called "*Reductio ad absurdum.*"

Exercise. — Use the above figure to establish the fact that the least distance from any point to a straight line will be the segment AF of the perpendicular.

This last distance named is the one always to be understood as the distance of a point from a line, or as the distance of a line from a point.

CHAPTER II.

A Triangle.

15. Definitions. The figure formed by three straight lines which intersect so as to enclose a portion of a plane is called a **triangle** ($\triangle$).

The points of intersection of the lines forming the triangle are called the **vertices**.

At each vertex four angles are formed. The angles within the enclosure are called **interior**.

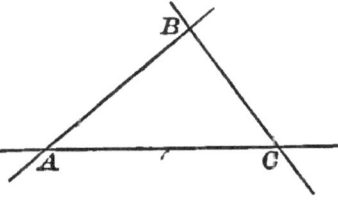

Fig. 15.

At each vertex there is an angle vertical to the interior one, and equal to it. There is also at each vertex a pair of vertical angles, each supplementary to the interior angle.

The segments AB, BC, and CA are called sides of the angle. Taken together they are called the **perimeter**.

Note. — The figure called a *triangle* might with equal propriety be called a *trilateral*.

16. If a point from a position (S) on the side AB, move in the directions indicated by the arrow-heads, and return to S after having moved along the perimeter of the triangle, it will have changed direction at B, at C, and at A; and not at any other point. At B it would change direction to the *left*, an amount indicated by $\angle 1$;

at C, it would change direction to the *left*, an amount indicated by ∠ 2, and at A it would change direction to the *left*, an amount indicated by ∠ 3.

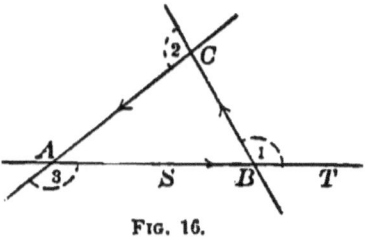

Fig. 16.

This appears to be a total change of direction equal to a complete rotation.

If it be such, we should be able to rotate some segment of a straight line about one of its extremities and have its partial rotations coincide with the changes of direction, and its complete rotation be the sum of the three changes of direction.

Let (ST) be any segment of one of the lines forming the triangle. Move it in the line, of which it is a part, until the point S coincides with B. Rotate the segment until it shall coincide in direction with BC. Then move it in the line BC until the point S comes to C. Then rotate until the direction CA is taken up. Along this line translate the segment until the point S shall coincide with A; then rotate until the direction AB is reached. Along this line move the assumed segment until it shall have come to its initial position.

The motion of the segment has been of two kinds: translation and rotation. The sum of the translations has been the perimeter of the triangle, and the sum of the rotations has been a complete rotation (360° or 4 rt. ∠s). Hence the

THEOREM. *The sum of the changes of direction (called exterior angles) equals 4 right angles.*

NOTES. — 1. If from the point S we should proceed in the opposite direction, the changes of direction at the points A, C, and B

would be *right* handed, and would be the verticals of angles 3, 2, and 1; and the sum would again be 4 right angles.

2. The student will observe that from a single triangle we have determined a relation which we say is true of all triangles. The reason for this is: we have drawn our conclusions without putting any limitations upon the triangle. *Any* triangle might have the same method applied to it with exactly the same result. We are then not drawing general conclusions from special cases.

Exercises. — 1. Show that the sum of the interior angles of a triangle $= 180°$.

2. Show that any angle of a triangle will lie between 0^c and $180°$.

3. Show that but one angle of a triangle may be $90°$ or greater.

4. Show that the sum of two sides of a triangle is greater than the third.

5. Given two angles of a triangle to find the third.

17. Theorem. *If two straight lines make equal angles with a third straight line intersecting them, they will make equal angles with any straight line intersecting them.*

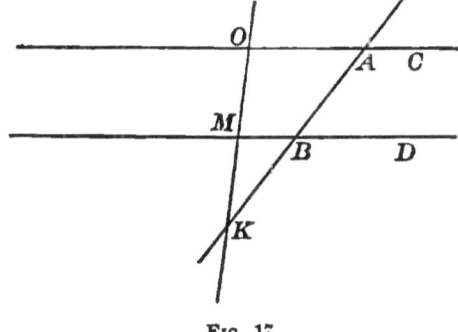

Fig. 17.

Let AC and BD represent the two lines, BA the third line, and MO any other line; and let the $\angle s$ CAK and DBK be the angles, that by hypothesis are equal.

$$\angle KAO + \angle AKO + \angle KOA = 180° \qquad \text{Ex. 1. § 16.}$$
$$\underline{\angle KBM + \angle BKM + \angle KMB = 180° \qquad \text{Ex. 1. § 16.}}$$
$$0 \;+\; 0 \;+ \angle KOA - \angle KMB = 0, \text{ by subtraction.}$$
$$\therefore \angle KOA = \angle KMB, \qquad \text{Axiom (c) § 12.}$$

Q. E. D.

Exercises. — 1. Establish the theorem when the fourth line passes through B.

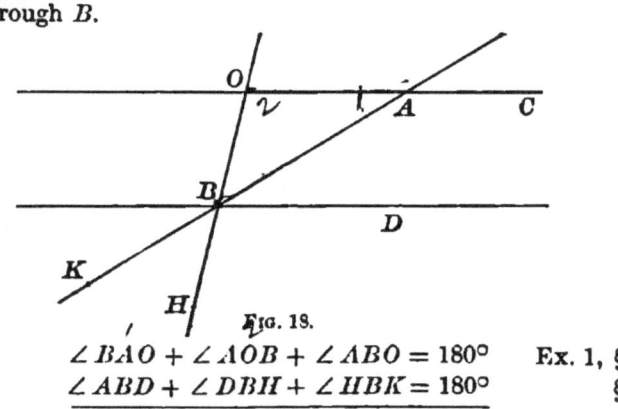

Fig. 19.

Proof. $\quad \angle BAO + \angle AOB + \angle ABO = 180°$ Ex. 1, § 16.
$$\underline{\angle ABD + \angle DBH + \angle HBK = 180° \qquad \text{§ 11.}}$$
$$0 + \angle AOB - \angle DBH + 0 = 0$$
By subtraction, $\quad \therefore \angle AOB = \angle DBH \qquad$ Axiom (c), § 12.

Q. E. D.

2. Establish the theorem when the fourth line does not intersect the third within the limits of the drawing.

Hint. — Draw an auxiliary line connecting two points of intersection and then apply Ex. 1. Two applications will be necessary.

3. Establish the theorem when the third and fourth lines intersect between the first and second.

NOTE. — In order to thoroughly familiarize himself with the methods of procedure and with the results, the student should make for himself figures that differ from those in the text both in relative proportions and in lettering, and use them to make demonstrations. He should be accurate and complete in enunciation and in demonstration, quoting his authority for each statement. Remember that neatness and accuracy of expression, whether written or oral, tend to accuracy of thought.

PARALLELS.

18. Definition. If two straight lines in a plane make equal angles with a third straight line in the plane, the two are said to be **parallel** (||).

The third is called a **secant**, or sometimes a **transversal**.

By § 17, parallels make equal angles with any secant.

AC and BD are said to have the same direction. Their opposites AF and BG have the same direction.

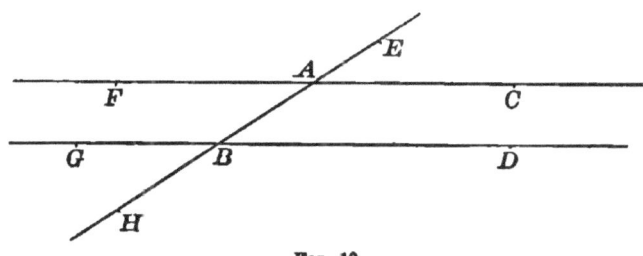

FIG. 19.

Exercises. — 1. Show that the eight angles about A and B are in two sets; those in one set being equal to each other, and those in the other set being equal to each other.

2. Show that the ∠ CAB and DBA are supplementary. Show the same of FAB and ABG.

19. Theorems. (a) *If the two lines which form an angle are parallel to the two lines which form another angle, the angles will be equal, if the lines forming one angle extend from its vertex in the same direction as the lines forming the other angle.*

(b) *If the lines forming one angle extend from its vertex in the opposite direction to the lines forming the other angle, the angles will be equal.*

22 ELEMENTS OF GEOMETRY.

(c) *If one set of parallels extend in the same direction, and the other set in opposite directions from the vertices, the angles will be supplementary.*

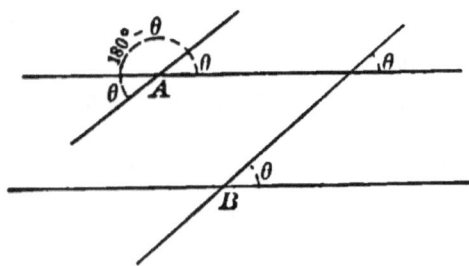

Fig. 20.

Let A and B represent the two vertices. The student will supply the demonstrations called for by these theorems. The figure furnishes suggestions.

Exercises. — 1. Show that if a line is perpendicular to one of two parallels, it is perpendicular to the other also.

2. Show that parallels are everywhere equally distant from each other.

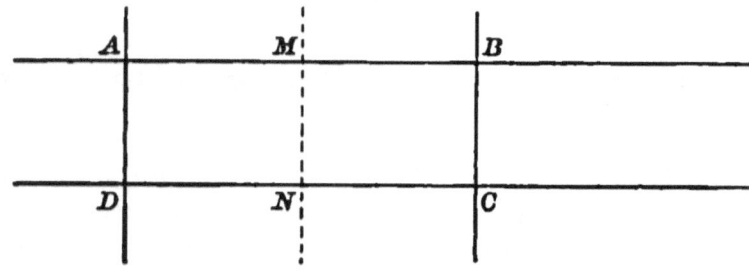

Fig. 21.

Proof. — *If* the segments AD and BC (which measure the distances between the parallels AB and DC at A and B) *are* equal, it is probable that they may be brought to coincide and their equality be thus established.

If at M, the middle point of the segment AB, an auxiliary line MN be drawn perpendicular to AB (§ 13) it will be perpendicular to DC (§ 19, Ex. 1). But we do not as yet know whether it will bisect DC or not.

If we revolve the portion of the figure that lies on the right of the line MN (§§ 8 and 10) about MN as an axis until the revolved portion coincides with the unrevolved portion, we shall have the segment MB coinciding with the segment MA (because M is by selection the middle point of the segment AB). The point B will fall at A, and the line BC (being perpendicular to MB) will in its revolved position take the direction AD.

Because NC is perpendicular to MN, it will when rotated lie in the direction ND.

The point C will then lie somewhere in the line AD, and somewhere in the line ND. It must lie at their intersection (D). Therefore the segment BC will coincide with the segment AD and be equal to it. Q. E. D.

3. Show that if two lines are parallel to a third line, they are parallel to each other.

4. Show how perpendiculars to a line may be drawn from points on the line. Let fall perpendiculars from given points to a given line. Through given points draw parallels to a given line.

Suggestion. — Use a ruler and a right-angled triangle; and assume the line in a variety of positions.

NOTE. — The formal statement of a *theorem* may be followed by the proof; or the relations leading to conclusions may be presented first and the formal statement of the theorem be presented at the end. When placed before the proof, it is a statement of relations *said* to exist. The proof is the establishing of these *said* relations (see § 17). When placed after the determination of relations it is in the form of a conclusion (see § 16).

In general it is better to state the theorem first, so that the student shall have in mind the relation that he is undertaking to establish.

20. Theorem. *If at the middle point of a segment of a line a perpendicular be erected, and if from any point in the perpendicular lines be drawn terminating at the extremities of the segment, they will be equal to each other.*

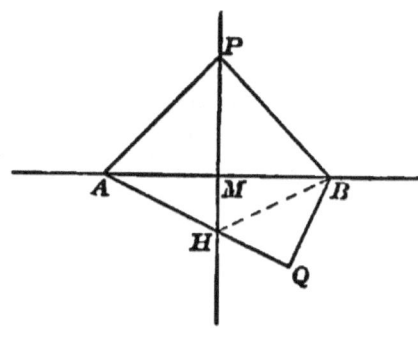

Fig. 22.

If $PB = PA$, it can be brought to coincide with PA, and since AB and MP are perpendicular to each other, a rotation of one part of the figure on a line as an axis is suggested.

Proof. Revolve the portion of the figure that lies to the right of the line MP, on MP as an axis, until it coincides with the plane on the left of MP. § 10.

The $\angle PMB$ will coincide with the $\angle PMA$. B will fall at A; P will remain stationary; and the segment PB will coincide with the segment PA, and must therefore be equal to it. Q. E. D.

Exercises. — 1. Show that any point not on the perpendicular bisector will not be equally distant from the extremities of the bisected segment.

2. Show that if a line have two of its points equally distant from the ends of the segment, it will be the ⊥ bisector of the segment.

Solution. — If a perpendicular bisector of the segment *were* erected it would contain *all* points that are equally distant from the extremities of the segment. For this reason it would pass through the *two* points mentioned in the hypothesis; but "*two* points determine the position of a straight line."

Therefore the line which by hypothesis passed through *two* points that were equally distant from the extremities of a given segment will coincide with and will *be* the perpendicular bisector of the given segment. Q. E. D.

NOTE. — The perpendicular bisector of the segment of a line is said to be the *locus* (place) of the point, when *moving* so that its distances from the segment ends shall always be equal to each other.

Or it may be described as the *locus* of a point *moving* so that the ratio of its distances from two fixed points always equals unity.

A CIRCLE.

21. Definitions. The locus of a moving point, the distance of which from a given point is fixed, is called a circumference.

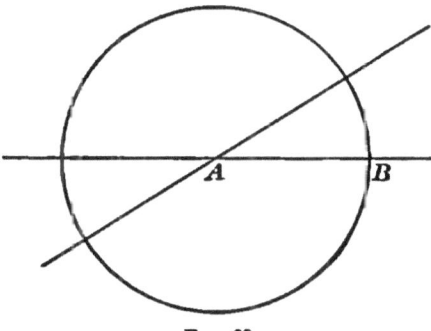

FIG. 23.

If the line AB rotate about A as a pivot, *any* point in the line AB, as the point B, in a complete rotation, will remain at a fixed distance from A, and will generate a circumference.

26   ELEMENTS OF GEOMETRY.

Anything less than a complete rotation will generate an arc. A half rotation will generate a semi-circumference; and a quarter rotation will generate a quadrant.

The point A is called the **centre**; and the distance of B from A is called the **radius**.

A point nearer to A than B is, will generate a circumference that will lie entirely within the circumference generated by the point B; and a point at a greater distance from A than B is, will generate a circumference, lying entirely outside of the circumference generated by B.

When AB rotates about A as a pivot, the change of direction makes an angle at A; each point of AB generates a circumference, and the whole line generates the surface of the plane.

The figure bounded by a circumference is called a **circle**.

CONGRUENT TRIANGLES.

22. Definition. When a geometric figure may be substituted for another and may be made to occupy exactly

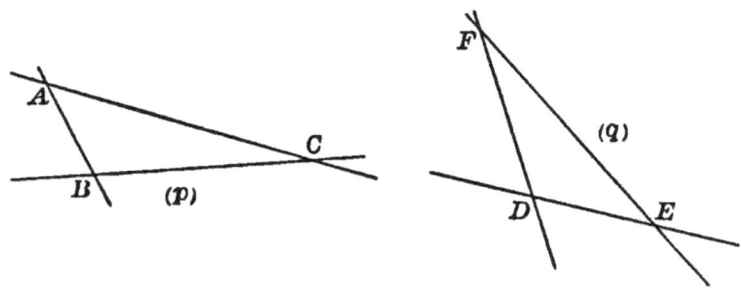

Fig. 24.

the same space which the other did, the figures are said to be *congruent*.

TRIANGLES.

THEOREM. *If two triangles have the three sides of the one equal to the three sides of the other, each to each, they are congruent.*

If the two triangles (p) and (q) having:
$$AB = DE,$$
$$BC = DF,$$
and $$AC = EF,$$

be so placed that a pair of equal sides shall lie together, and the triangles be not superimposed, we shall have them placed as in figure (r).

By the terms of the theorem, B and C are two points equally distant from A and E.

Draw the auxiliary line AE.

By § 20, Ex. 2, the line CB will be the perpendicular bisector of the segment AE.

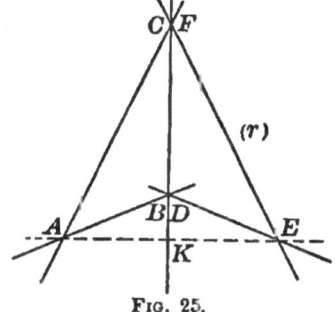

Fig. 25.

If the figures to the right of the line CK be revolved on CK as an axis, the point E will fall at A; the segments ED and AB will coincide; the same will be true of EF and AC; DF and BC will remain in coincidence; and *all* the parts of one triangle (perimeter, angles, and surface) will coincide with the parts of the other. Q. E. D.

23. THEOREM. *If two triangles have two angles and an included side of one equal to two angles and an included side of the other, they are congruent.*

The student will make the necessary constructions, and show that the triangles may be placed so as to coincide.

Exercise. — Show that two triangles which have any two angles and a side of one equal to two angles and the corresponding side of the other are congruent.

NOTE. — *Corresponding* side means one that is placed in like relation to the equal angles.

Corresponding sides are often called *homologous* sides.

24. THEOREM. *If two triangles have two sides and the included angle of one equal to two sides and the included angle of the other, they are congruent.*

The student will supply the necessary constructions, and make application of one figure to the other.

Exercise. — Show what significant position two triangles may be made to occupy with respect to each other, when two sides of one are equal to two sides of the other and the included angles are supplementary.

CONSTRUCTION OF TRIANGLES.

Four Cases.

25. FIRST CASE. When three sides are given.

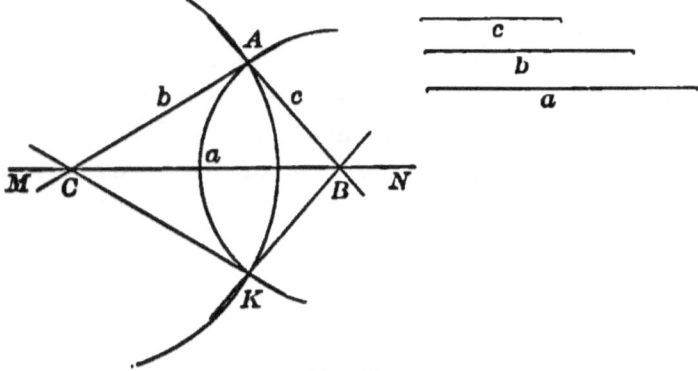

FIG. 26.

On the line MN lay off a segment (a) with ruler, tape, or compasses. C and B will mark two of the vertices of the required triangle. With C as a centre and a radius (b) construct an arc; and with B as a centre and radius (c) construct another arc. From the intersection A, draw the lines AC and AB.

NOTE. — It is customary, as a matter of convenience, to designate the vertices of a triangle by the capital letters, and the sides *opposite* to them by the same letters in small type.

The triangle, the vertices of which are A, B, and C ($\triangle ABC$), will be the required triangle. Q. E. F.

The construction can be made when the sum of any two of the segments, given as sides, is greater than ($>$) the third side; but not otherwise.

The student will substantiate this statement.

Exercises. — 1. Show that if two circumferences intersect in one point, they will intersect in two points.

2. Show that the same will be true if a straight line intersects a circumference.

3. Show that in the figure the $\triangle CKB$ is congruent with the $\triangle CAB$.

4. Show that two other triangles might be constructed by interchanging the radii (b) and (c); and that the four triangles thus constructed would be congruent.

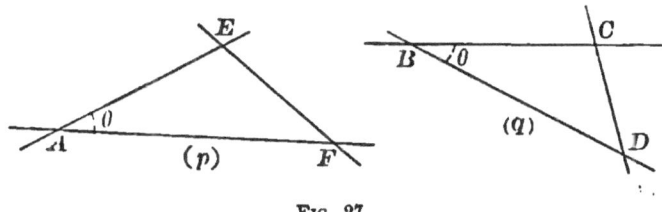

FIG. 27.

5. With ruler and compass show how to construct, in any position, an angle that shall be equal to a given angle.

Suggestion. — Let θ (Fig. p) represent the given angle. If we draw any line (EF), intersecting the lines forming the angle, so as to obtain a triangle having θ for one of its angles, we shall have a triangle, the *three* sides of which may be used to construct a congruent triangle in any desired position, as in (Fig. *q*).

It is customary, as a matter of convenience, to assume AF and AE equal to each other.

26. SECOND CASE. That in which two angles and the included side are given.

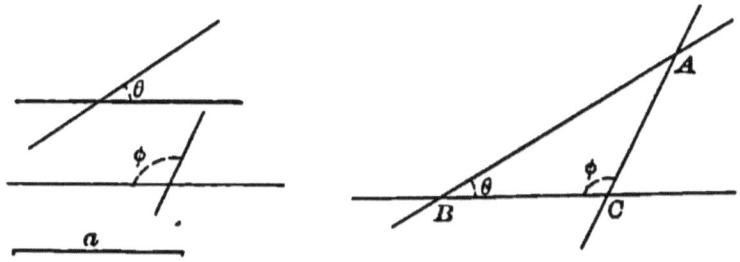

Fig. 28.

On the indefinite line BC lay off the segment BC equal to the given side, and at B and C construct the given angles, so that they shall be interior.

There will be two constructions, but the two will be congruent and may be considered as one.

NOTE. — This case may be considered as including the one in which are given two angles and a side opposite one of them; for if two angles are given, the third may be determined, and the method of this case may then be applied.

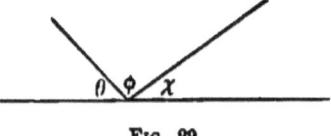

Fig. 29.

Exercises. — 1. Make a construction when $a = 3$ inches, $\theta = 80°$, and $\phi = 120°$.

2. Show how to construct a triangle, if we have given: a side, an angle adjacent, and an angle opposite.

27. THIRD CASE. That in which two sides and the included angle are given. Construct the angle; from the vertex lay off the given sides; and join their extremities.

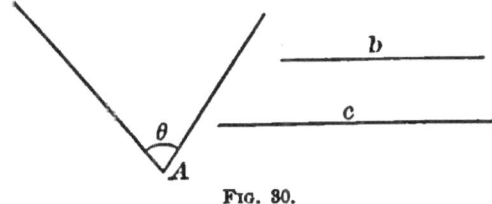

FIG. 30.

The student will show that there are two constructions, but the resulting triangles are congruent.

28. FOURTH CASE. That in which two sides and an angle opposite one of them are given.

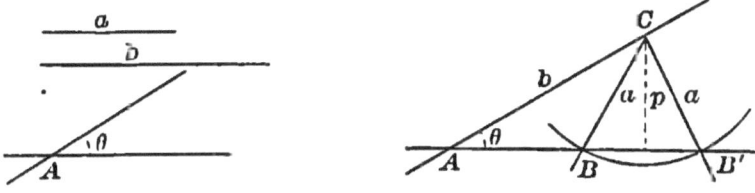

FIG. 31.

Construct an angle equal to θ. On either side of the angle lay off the segment (b). Let C be its other extremity. From C as a centre with a radius (a) construct an arc. Join C with the points in which the arc intersects the other side of the angle. (See § 25, Ex. 2.)

Exercises. — 1. Show that if (a) be less than the $\perp$ (p), there cannot be any construction.

2. Show that if $a = p$, there can be one construction, provided $\theta < 90°$, and not any if $\theta \gtreqless 90°$.

3. Show that if a be intermediate in value between b and p, ($b > a > p$), there can be *two* constructions if $\theta < 90°$, and *none* if $\theta > 90°$.

4. Show that if $a = b$, there can be one construction when $\theta < 90°$, and none when $\theta > 90°$.

5. Show that if $a > b$, there can be one construction when $\theta < 90°$, and one when $\theta > 90°$.

6. Show that if the three angles are given, the triangle will not be determined.

NOTES. — 1. When two or more constructions (not congruent) are possible, the case is said to be ambiguous.

2. In order to construct a plane triangle, three parts must be given, one of which is a line.

GENERAL EXERCISES.

1. Show that lines which are not parallel will intersect.

2. Show how to bisect a given segment of a line.

3. Show that oblique lines, from a point in a perpendicular, intersecting the base line at different distances from the foot of the perpendicular, make unequal angles with the base line.

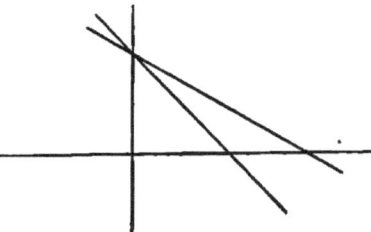

FIG. 32.

4. Show that an exterior angle of a triangle equals the sum of the interior non-adjacent angles.

INEQUALITY AXIOMS.
(a) *If the greater members of two inequalities be added, the sum will be greater than the sum of the lesser members.*

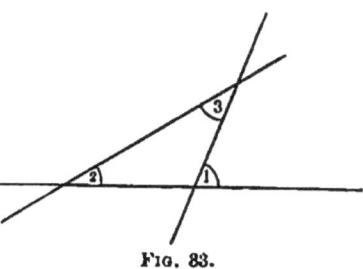

FIG. 33.

(b) *If equals be added to or subtracted from the two members of an inequality, the inequality will still exist, and will exist in the same sense.*

(c) *If the members of an inequality be multiplied or be divided by a positive quantity, the inequality will subsist in the same sense; but if multiplied or divided by a negative quantity, the inequality will be reversed.*

5. Show that the sum of two sides of a triangle is greater than the sum of the distances from any point within the triangle to the extremities of the third side.

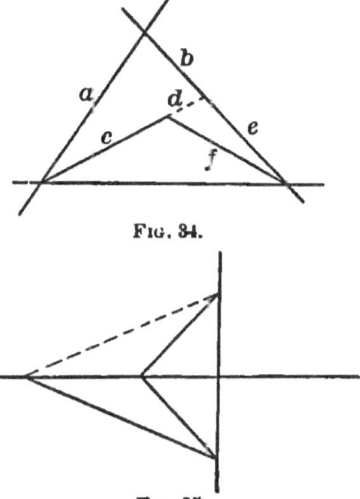

Fig. 34.

$$a + b > c + d$$
$$d + e > f$$
$$\overline{a + b + e > c + f}$$

6. Show that if from any point in a perpendicular, oblique lines be drawn to the base of any two, the one which meets the base at the greater distance from the foot of the perpendicular will be the greater.

Fig. 35.

7. Show how to construct a triangle, having given two angles and the side opposite one of them without having found the third angle.

8. Show that the difference of two sides of a triangle is less than the third side.

9. Through a point to draw a line parallel to a given line, and show that only *one* such line can be drawn.

10. Show how we may, with ruler and compass, erect a perpendicular at any point of a line, and how we may let fall a perpendicular from a point.

11. Prove the converse of Ex. 2, § 18.

NOTE. — A proposition is a statement of a relation *said* to exist, and is in the general form of subject and predicate.

The *converse* of a proposition is also a proposition; but with relations of subject and predicate reversed.

Exercise 2, § 18, as a formal proposition would be: (If the two lines AC and BD are parallel and are intersected by a third line HE)(then will)(the angles BAC and ABD be supplementary). The converse would be: (If two lines are met by a third line so as to make the interior angles on one side of the secant supplementary)(then will)(the two lines so situated with respect to the third be parallel).

The *Reductio ad Absurdum* method is particularly well adapted to the determination of the truth or falsity of the converse of a proposition.

CHAPTER III.

29. Definitions. A triangle is called:

Right, when one of its angles is 90°.

Oblique, when none of its angles are 90°.

Obtuse, when one of its angles is $> 90°$.

Acute, when each of its angles is $< 90°$.

Equiangular, when the three angles are equal to each other.

Equilateral, when the three sides are equal to each other.

Isosceles, when two sides are equal to each other and not equal to the third side.

Scalene, when there is not any equality between sides.

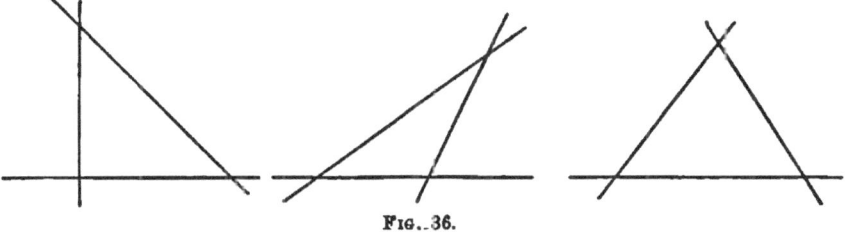

Fig. 36.

Special names given to parts of triangles are:

Base, which may be any one side.

Base angles, which are the angles adjacent to the base.

Vertex, which is the point of intersection of the sides, not considered the base.

Vertex angle, which is the angle of the triangle at the vertex.

Hypothenuse, which is the side opposite a right angle.

Altitude, which is the perpendicular from vertex to base.

30. Theorem. *If at the middle of a side of an equilateral triangle a perpendicular be erected, it will pass through the opposite vertex.*

If at M, the middle point of the side AB, a perpendicular be erected, it will contain all points that are equally distant from A and B. C is such a point. Q. E. D.

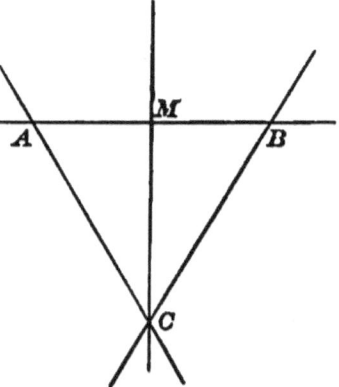

Fig. 87.

Exercises.—1. Show that the angle at the vertex will be bisected.

2. Show that if perpendiculars be let fall from the three vertices to the opposite sides of an equilateral triangle, they will be equal to each other.

3. Establish the converse of the theorem.

4. Show that an equilateral triangle is equiangular.

5. Establish the converse.

31. Theorem. *If a perpendicular be erected at the middle point of the non-equal side of an isosceles triangle, it will pass through the opposite vertex.*

The proof is the same as in § 30.

Exercises.—1. Show that the angles opposite the equal sides are equal.

2. Show that the angle at the vertex is bisected by the perpendicular.

3. Show that if perpendiculars be let fall from the vertices of the equal angles, they will be equal to each other.

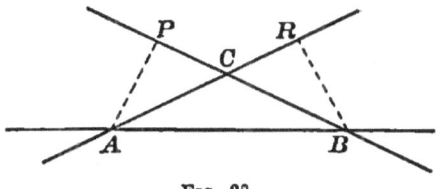

FIG. 38.

Solution. — Let ACB represent the isosceles triangle; and AP and BR the perpendiculars.

If the perpendiculars *are* equal, the △ APC and BRC will be equal because they will then have a perpendicular and an hypothenuse of a right triangle the same in each.

Does this relation (the equality of the triangles) exist without the consideration of the perpendicular?

In the △ APC and BRC,

$$AC = BC, \qquad \text{(by hypothesis)}$$
$$\angle APC = \angle BRC, \qquad \text{(being } 90°\text{)}$$
$$\angle ACP = \angle BCR, \qquad \text{(being vertical)}$$

Hence by (§ 23, Ex.), $\triangle APC = \triangle BPC$.

AP and BR are corresponding parts and are equal.

The question has thus been answered: The necessary relation does exist, and the problem is solved. Q. E. D.

4. Solve the same problem, using for the figure an acute-angled triangle.

5. Show that if two angles of a triangle are equal, the sides opposite those angles are equal, *i.e.* the triangle will be isosceles.

NOTE. — A problem is something proposed to be done. The solution is the finding of sufficient previously established relations to warrant the doing.

32. Theorem. *If two angles of a triangle are unequal, the sides opposite them are unequal, and the greater side lies opposite the greater angle.*

Let ABC represent the triangle, the angles of which at A and B are unequal.

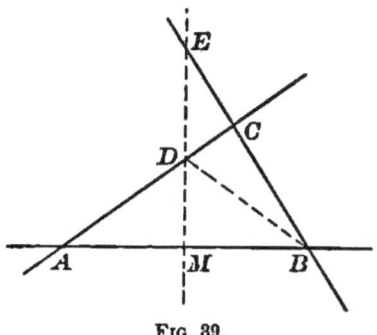

Fig. 39.

If at the middle point (M) of the side AB a perpendicular be erected, it could not pass through C; for if it did, the triangle would be isosceles. Not passing through C, it must intersect the lines forming the other sides of the triangle at separate points. Let D be the intersection that is the nearer to M.

Draw the auxiliary line DB. It will separate the angle to which it is drawn into two parts, one of which equals the $\angle CAB$. The $\angle B$ is the larger.

$$AC = AD + DC = BD + DC > BC. \qquad \text{Q. E. D.}$$

Exercises. — 1. Prove the same by using $\overline{AE}$ as an auxiliary line.

2. Show that the hypothenuse of a right triangle is greater than either of the other sides.

3. Establish the theorem of this section, with an obtuse-angled triangle.

4. Establish the converse of the theorem.

5. Show that if two right triangles have one of the sides adjacent to the right angle and the hypothenuse mutually equal, they are congruent.

TRIANGLES.   39

33. Theorem. *Any point of an angle bisector is equally distant from the lines forming the angle.*

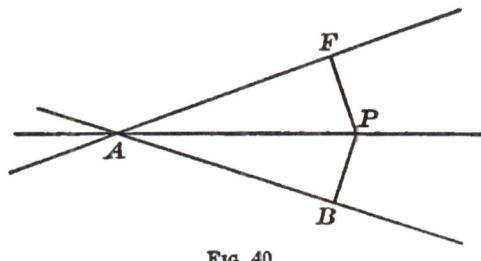

Fig. 40.

If AP represent the angle bisector, and if from any point (P) perpendiculars be let fall to the lines forming the angle, the △ PAF and PAB will be congruent, having two angles and the included side of one equal to two angles and the included side of the other.

Hence $\qquad PF = PB.$ \hfill Q. E. D.

Theorem. *The distances of any point, not on the angle bisector, from the angle lines will not be equal.*

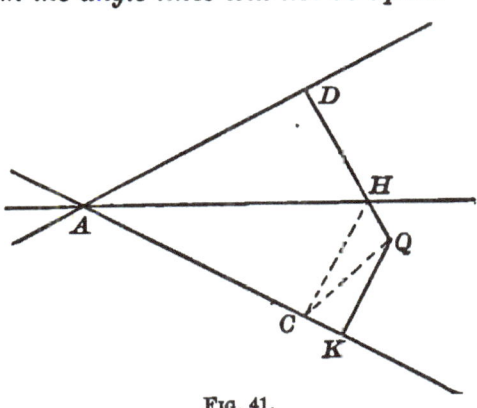

Fig. 41.

Let Q represent *any* point not on the angle bisector. Let QD and QK be the perpendiculars. One of them intersects the angle bisector. Let *H* be the point of such intersection.

Draw $HC \perp$ to AK, and join QC.
$$QD = QH + HD = QH + HC > QC > QK.$$
$$\therefore QD > QK.$$
Q. E. D.

Exercise. — Show that the bisector of the supplement of $\angle KAD$ will be perpendicular to AH, and that every point in it will be equidistant from the angle lines.

Note. — If perpendiculars on one side of a line are positive (+), those on the other side are negative (−).

The locus of a point, the ratio of the distances of which from two lines is (+ 1), will be the angle bisector If the ratio of the distances is (− 1), the locus will be the bisector of the supplement of the first angle and will be perpendicular to the bisector of the first angle.

34. Problem. *To construct a bisector of an angle.*

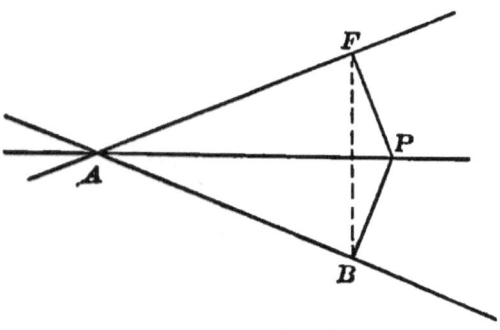

Fig. 42.

If *AP* were the angle bisector required, and *if* we should draw the perpendiculars *PB* and *PF*, they would be equal to each other, and the segments *AB* and *AF* would be equal. If the auxiliary line *BF* were drawn, *AP* would be a perpendicular bisector to it.

TRIANGLES. 41

This determination of relations that *would* exist *if* the angle bisector *were* drawn suggests the construction.

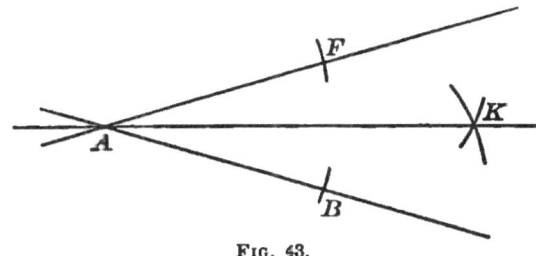

Fig. 43.

Lay off AB equal to AF, and find some point (K), other than (A), which is equally distant from B and F.

Draw AK; it will be the angle bisector; since it is the perpendicular bisector of $\overline{BF}$.

Note. — We might have followed more closely the relations at first developed, by erecting at B and F perpendiculars to $\overline{AB}$ and $\overline{AF}$, and have joined their intersection with A.

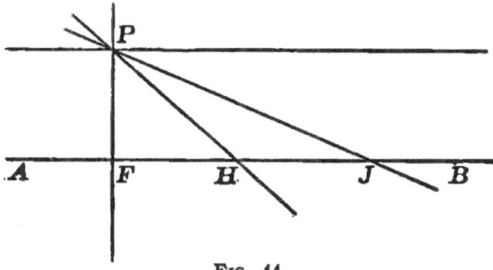

Fig. 44.

35. If through the point P a perpendicular to the line AB be drawn, and then be rotated positively (*i.e.* to the *left*), the point in which it intersects AB will move from F in the direction of B. When it shall have rotated 90°, the line through P will be parallel to AB and will not intersect it. We mean the same thing when we say that the point of intersection has passed to infinity.

The instant the angle exceeds 90°, the rotating line again intersects the line AB, in the direction FA. The distance from F will continually diminish until 180° of rotation has taken place, when the point of intersection will have reached F.

The next 180° the point of intersection will travel over the same route as before.

QUADRILATERALS AND QUADRANGLES.

36. A plane figure formed by four straight lines which enclose an area forms a **quadrilateral**. If each line inter-

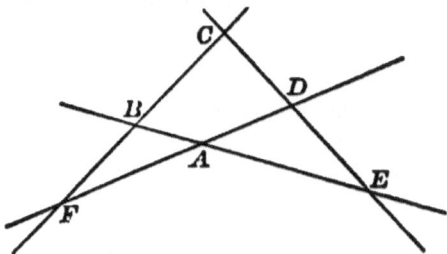

FIG. 45. — QUADRILATERAL.

sects each of the others as indicated in the above figure, the figure is called a **complete** quadrilateral.

The portion $ABCD$ is called a **quadrangle**, the vertices of which are at A, B, C, and D, and the sides of which are $\overline{AB}$, $\overline{BC}$, $\overline{CD}$, and $\overline{DA}$.

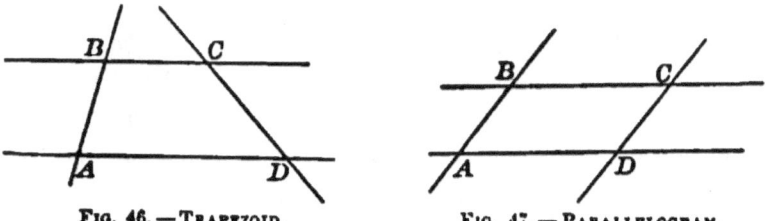

FIG. 46. — TRAPEZOID.   FIG. 47. — PARALLELOGRAM.

If two sides are parallel, the quadrangle is called a **trapezoid**.

QUADRILATERALS AND QUADRANGLES. 43

If the sides of the quadrangle are parallel two and two, the figure is called a **parallelogram**.

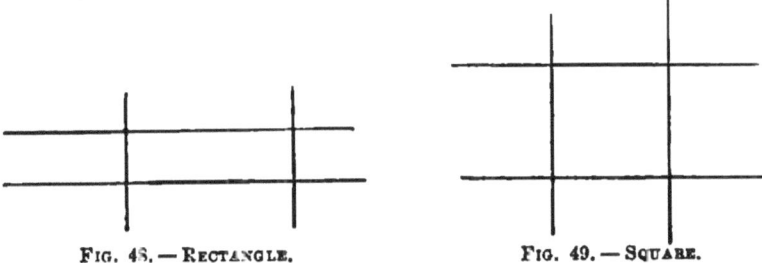

Fig. 48.— Rectangle.   Fig. 49.— Square.

If a parallelogram be right-angled, the figure is called a **rectangle**.

If the rectangle have all its sides equal to each other, the figure is called a **square**.

If the sides of an oblique parallelogram (none of the angles being 90°) are equal,

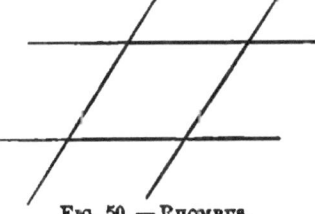

Fig. 50.— Rhombus.

the figure is called a **rhombus**. A quadrangle not having any side parallel to any other side is sometimes called a **trapezium**.

37. Theorem. *The sum of the exterior angles of a quadrangle is 360°, or four right angles.*

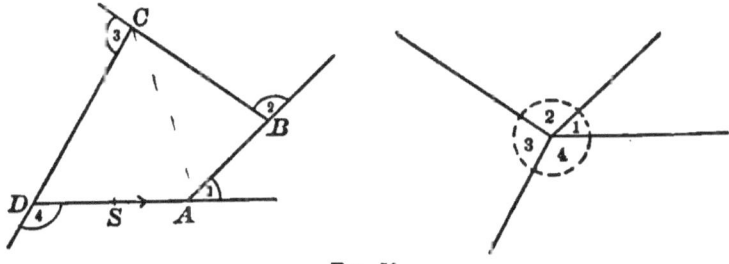

Fig. 51.

A point S moving in the direction indicated, about the perimeter of the quadrangle and arriving at its initial position, will have made *left*-handed changes of direction at the four vertices. These changes of direction being placed adjacent to each other, as indicated in the second figure, give a complete rotation, or 360°. Q. E. D.

Exercise. — Show that the sum of the interior angles is *four right angles*.

38. Theorem. *The opposite sides of a parallelogram are equal.*

If the theorem be true, and we draw an auxiliary line connecting a pair of non-adjacent vertices (such a line is called a *diagonal*), we would have the figure separated

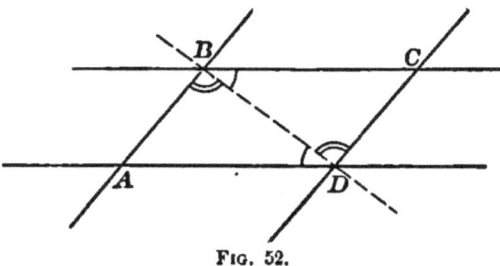

Fig. 52.

into two triangles, which would have the three sides of one equal to the three sides of the other, and would be equal.

Can we show that the triangles are equal (congruent), making use only of previously established relations?

We can, because:
$$\angle CBD = \angle ADB,$$
$$\angle CDB = \angle ABD,$$
and $$\overline{BD} = \overline{BD}.$$

The two triangles have two angles and the included side in each equal, and are therefore congruent (§ 23); and hence the theorem.

Exercises. — 1. Use the theorem to show that parallels are everywhere equally distant from each other.

2. If a pair of non-adjacent sides in a quadrangle are parallel and equal, show that the figure will be a parallelogram (▱).

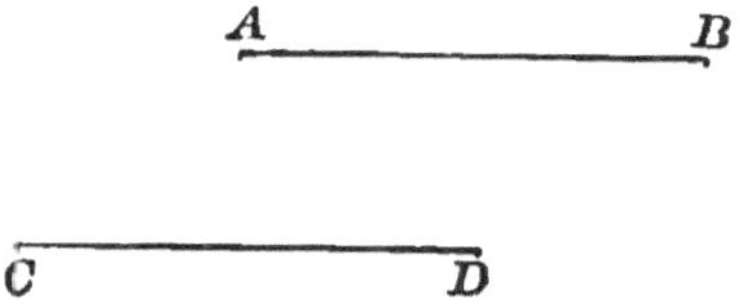

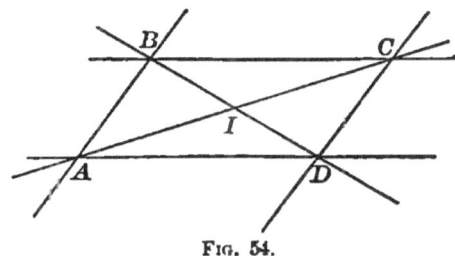

Fig. 53.

39. Theorem. *The diagonals of a parallelogram mutually bisect each other.*

Fig. 54.

If the diagonals do bisect each other, the △ BIC and AID will be equal.

Are they?

They are, because:

$$\angle IBC = \angle IDA,$$
$$\angle ICB = \angle IAD,$$
and $$\overline{BC} = \overline{AD}.$$
From which $$\triangle BIC = \triangle AID.$$
$$\therefore BI = ID \text{ and } AI = IC.$$

Q. E. D.

Exercises. — 1. Show that the diagonals of a rectangle are equal to each other.

2. Show that the diagonals of a rhombus are perpendicular to each other.

3. Show that the diagonals of a square are equal, and are perpendicular to each other.

40. Definitions. A **polygon** is a figure formed by a number of straight lines which enclose an area.

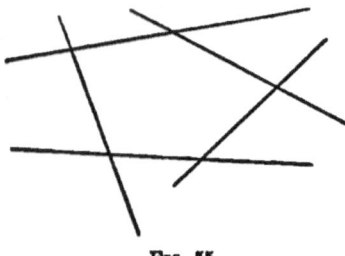

Fig. 55.

In a polygon there are as many angles as sides.

A Triangle	has 3 angles.	A Nonagon	has 9 angles.
A Quadrangle	has 4 angles.	A Decagon	has 10 angles.
A Pentagon	has 5 angles.	An Undecagon	has 11 angles.
A Hexagon	has 6 angles.	A Dodecagon	has 12 angles.
A Heptagon	has 7 angles.	A Pendecagon	has 15 angles.
An Octagon	has 8 angles.		Etc.

If all the angles are equal to each other, and all the sides are equal to each other, the polygon is *regular*.

An *exterior* angle of a polygon is the change of direction in going from one side to an adjacent side.

Any polygon in which all the changes of direction are in the same sense, is called a **convex** polygon.

If the changes of direction are not all in the same sense, the polygon is said to be *re-entrant*.

POLYGONS. 47

41. Theorem. *The sum of the exterior angles of any polygon is* 360°.

If on the perimeter of any convex polygon as represented in the accompanying figure we take any point as (S), and traverse the perimeter, starting in the direction indicated, and returning to (S), we shall at the vertices, A, B, C, D, and E, have made changes of direction to the left, amounting in all to a complete rotation, or 360°.

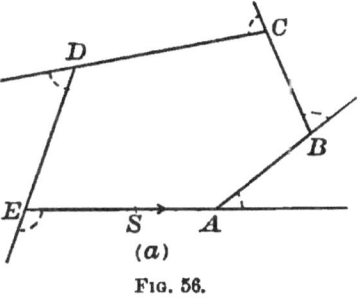

Fig. 56.

Figure (b) represents a re-entrant polygon.

As in the convex polygon, traversing the perimeter from (S), starting in the direction indicated and arriving

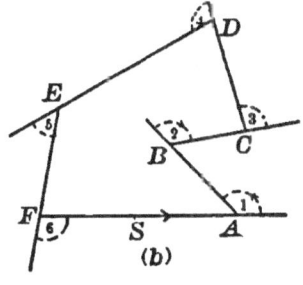 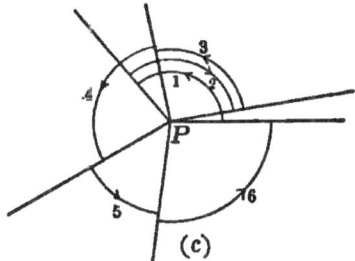

Fig. 57.

at (S), a complete rotation will have been made. But it is to be noted that at A the change of direction is to the *left*, at B it is to the *right*, at C, D, E, and F it is to the *left*.

Let P (Fig. c) be any point in the plane. From P draw lines parallel to the lines of Fig. b, which indicate the changes of direction at the succeeding vertices.

48 ELEMENTS OF GEOMETRY.

We see that the angles at A, C, D, E, and F are positive, while the angle at B is negative. And we see that the aggregate, or the algebraic sum, is 360°. Q. E. D.

NOTE. — The proof is not in any way affected by the number of sides of the polygon or by the number of re-entrant angles. We are then entitled to draw the conclusion we have.

An *interior* angle has been defined as the supplement of the corresponding exterior angle. An *exterior* angle has been defined as the change of direction at a vertex. Then the interior angle at B would be $180 - (-\theta) = 180° + \theta$; an appropriate value, as may be seen by drawing auxiliary lines from B to D, E, and F; and then taking the sum of the interior angles of the triangles thus formed.

Exercises. — 1. Find the sum of the interior angles of a pentagon. Find what each one will be if the pentagon be regular.

2. Do the same for polygons of 6, 7, 8, 9, 10, 11, and 12 sides.

ANALYSIS.

The Way to attack a Problem.

42. In every field of investigation problems are presented. These problems require solution. The solving of a problem is the determination of sufficient, previously established relations, to warrant the conclusion which is said to exist, or which appears to exist, or which is desired to exist.

The manner of approaching the solution of any problem is the same in all subjects, i.e. *we are to approach it through the analysis.*

When one makes an analysis, he asks himself one of the three following questions:

IF the relations $\begin{Bmatrix} said \\ which\ appear \\ desired \end{Bmatrix}$ to exist, do exist, what are the necessary, previously established, and sufficient relations ?

If the *necessary, previously established,* and *sufficient* relations are found and applied, the problem is solved.

NOTE. — Students frequently concern themselves with problems which they have no business to attack, for the reason that their information is not sufficient. When one attempts the making of the analysis, if the problem be an inappropriate one for him to attempt, he will presently become aware of that fact; and further time need not be wasted, for unless *sufficient necessary* relations have been previously established by him, there is no power to solve the problem.

The most important thing in education is the learning to make an analysis of every problem which we desire to solve. The instant that one asks himself the question formulated above, he puts himself in the proper frame of mind for determining whether the *stated, apparent,* or *desired* relations have sufficient foundation. It emphasizes the necessity of having facts and relations developed in their proper order, each with sufficient basis.

GENERAL EXERCISES.

1. If two angles can be so placed as to have a common side and other two sides intersect, the sum of the intersecting sides will be greater than the sum of the non-intersecting sides.

2. Show that if from any point within a triangle, lines be drawn to two vertices, the difference of the interior angles at P and C will equal the sum of the △ PAC and PBC.

3. Show that if two triangles have two sides of one equal to two sides of the other, and their included angles unequal, the triangle having the greater included angle will have the greater third side.

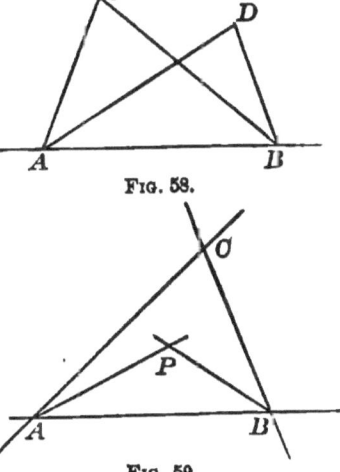

FIG. 58.

FIG. 59.

E

Suggestion. — The vertex C of the triangle having the smaller included angle ($\angle ABC < \angle ABD$) will, when a pair of equal sides are placed together, fall as indicated in *one* of the figures.

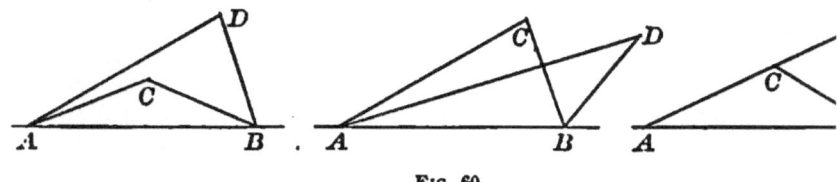

Fig. 60.

4. Show that if a right triangle have one of its oblique angles 30°, the side opposite the 30° angle will be one-half the hypotenuse.

5. Show which of the eight parts of a quadrangle, when given, will be sufficient to determine a construction.

6. Show that the bisectors of the interior angles of a triangle are concurrent (pass through one point).

Remark. — Be careful not to assume that they are concurrent.

CHAPTER IV.

CIRCLES.

43. Definitions. A **secant** of a curve is a straight line which intersects the curve. A secant will intersect a curve in two points. The nature of a curve may be such that a secant may intersect it in more than two points.

A **chord** is the segment of a secant between two points of intersection. The portion of the curve between the same points subtends the chord.

The **centre** of a curve is a point through which, if *any* straight line be drawn intersecting the curve, the chord will be bisected at that point.

A **diameter** is a chord which passes through the centre.

A **tangent** to a curve is a straight line having a point in common with the curve, and having the same direction that the generating point of the curve has, at the common point. The common point is called the point of **contact**.

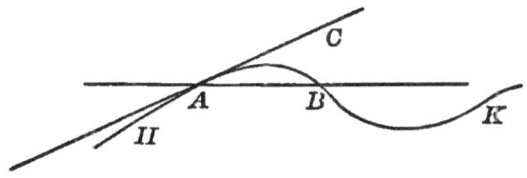

FIG. 61.

If we have a curve, as the one indicated by $HABK$, and if we cause any secant, as AB, to rotate about A as a pivot, so that B shall move along the curve toward

A, and eventually coincide with *A*, the secant will be a secant until *B* coincides with *A*, when it will be a tangent; for the straight line will at that instant have the same direction that a point in motion along the curve will have at *A*.

A *tangent* is sometimes said to be the limit toward which the secant approaches as the points of intersection approach coincidence.

A point being *position* and not having magnitude, *B* may pass through and beyond the position *A*. When that happens, the rotating line will have again become a secant.

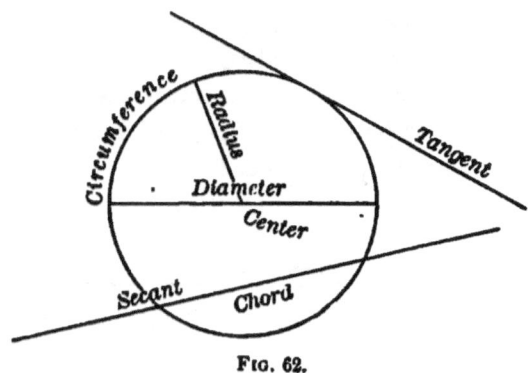

Fig. 62.

NOTE.— All curves have secants and chords; but comparatively a small number of curves have centres and diameters. The circle is one of this small number.

44. THEOREM. *A tangent to a circumference is perpendicular to the radius drawn to the point of contact.*

Analysis. — If a tangent to a circumference be perpendicular to a radius, the rotating line which was a secant and became a tangent must approach an angle of 90° with the radius to the point of rotation and arrive at 90° at the limit.

CIRCLES. 53

Proof. — Let AB be a tangent at A. Draw any secant as AM. The $\triangle ACM$ is isosceles, and the $\angle CAM$ is less than 90°. As the secant rotates about A as a pivot, and M approaches A, the $\angle ACM$ will approach 0, and the $\angle CAM$ will approach 90°.

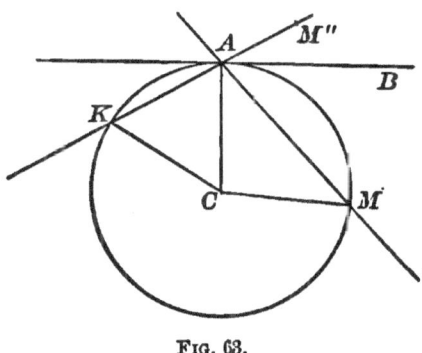

Fig. 63.

After M has passed through A, the angle CAM'' will be greater than 90°, because $\angle CAK$ is less than 90°.

As the change has been a continuous one, the angle made with CA changing gradually from an acute to an obtuse angle must have passed through 90°. And furthermore, when the second point is on either side of A the angle is oblique, it must be 90° when the second point coincides with the pivotal point, and the rotating line has become a tangent. Q. E. D.

45. Theorem. *The perpendicular bisector of a chord of a circle will pass through the centre and will bisect the arcs subtended by the chord.*

Section 20 furnishes the proof for the first part of the theorem.

For the establishing of the second part, the analysis suggests that we revolve one portion of the figure on PQ as an axis. $\overline{PQ}$ is the diameter of the circle that is the perpendicular bisector of $\overline{AB}$.

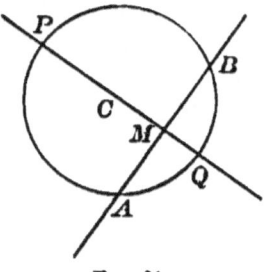

Fig. 64.

When revolved, MA will coincide with MB, and the point A will fall at B. The points Q and P will remain stationary.

The two circumferences will coincide, for every point of each will be at the same distance from C.

Hence $\overset{\frown}{QA}$ will coincide with $\overset{\frown}{QB}$, and $\overset{\frown}{PA}$ will coincide with $\overset{\frown}{PB}$.

Note. — The chord AB subtends the two arcs AQB and APB. But ordinarily, in speaking of the arc subtended by a chord, the lesser arc is the one understood.

Exercises. — 1. Show that chords AQ and BQ would be equal to each other; and that chords AP and BP would also be equal to each other.

2. Show how to draw a tangent at a given point of a circumference.

3. Having given a circumference, show how to find the centre.

46. Recalling the matter in §§ 7, 11, and 21, and again observing the generation of an angle by the rotation of a line about one of its points as a pivot, we are prepared to develop another relation; viz. the

Theorem. *If a circumference be constructed with the vertex of an angle as its centre, the arc included between the lines forming the angle will be the same fractional part of the entire circumference that the angle is of 360°.*

During a complete rotation every point in the line AB will generate a circumference. Let B represent one of these points.

If the angular magnitude about A be generated by a uniform rotation of the line AB, any point in the line AB will move at a uniform rate.

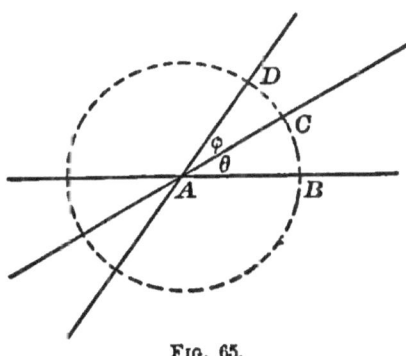

Fig. 65.

The angle and the circumference are generated with uniformity; they are begun at the same instant; and the 360° of rotation are completed at the instant the circumference is completed. At any stage of the proceeding, therefore, the angle generated will be the same fractional part of 360° that the arc generated by *any* point is of the entire circumference, or

$$\frac{\theta°}{360°} = \frac{\widehat{BC}}{\text{circum.}}$$

COROLLARY. *Two angles having their vertices at the centre of a circle will have the same ratio as the intercepted arcs.*

$$\frac{\theta°}{360°} = \frac{\widehat{BC}}{\text{circum.}} \tag{1}$$

and
$$\frac{\phi°}{360°} = \frac{\widehat{CD}}{\text{circum.}}. \qquad (2)$$

Dividing the members of (1) by the members of (2), we get

$$\frac{\theta°}{\phi°} = \frac{\widehat{BC}}{\widehat{CD}}.$$

Q. E. D.

NOTES. — 1. A *corollary* is a subsidiary theorem that follows from a principal one. In this work there are very few corollaries presented; it being preferred to establish the facts and relations as original exercises.

2. Because of the fact that when the vertex of an angle is at the centre of a circle, its sides intercept the same fractional part of the circumference that the angle is of 360°, we say that the angle is measured by the intercepted arc.

For purposes of numerical description, in speaking of a single angle when the angle itself is not represented in the drawing, an angle of 1° is applied as many times as it will be contained in the angle; then the *one sixtieth* part of one degree (or an angle of one minute) is applied to the remainder as many times as it will be contained in it; then to the second remainder is applied an angle of one second; and if a nearer approximation is desired, the decimal subdivisions of a second are applied to the succeeding remainders until the numerical description is as accurate as the circumstances demand.

In the same way, when numerical description is needed in order to represent an arc of a circumference, an arc which subtends an angle of 1° is applied as far as possible; to the first remainder is applied an arc that subtends an angle of 1', until there is a remainder less than 1'; to this is applied the arc that subtends 1'', etc.

Ordinary surveying instruments describe an angle to within 30''; very accurate geodetic instruments to within 10''; ordinary astronomical instruments to within 3''; and the best to within $\frac{1}{10}$''.

CIRCLES.

47. Definitions. If two chords intersect on the circumference of a circle, the angle they make is said to be an inscribed angle, and the intercepted arc is said to subtend the angle.

Theorem. *An inscribed angle is measured by half of the intercepted arc.*

(a) If one of the chords be a diameter as AB, draw the auxiliary line CD.

$$\overline{CA} = \overline{CD}.$$
$$\therefore \angle CAD = \angle CDA.$$
But $\angle BCD = \angle CAD + \angle CDA$
$$= 2 \angle CAD.$$

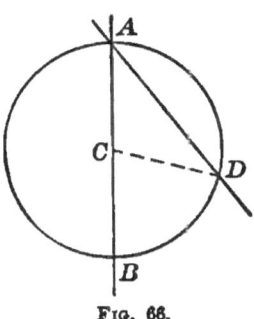

Fig. 66.

The $\angle BCD$ is measured by the arc BD. Hence the $\angle BAD$ is measured by one-half the arc BD.

(b) If the centre be within the angle formed by the two chords, draw an auxiliary diameter and then demonstrate.

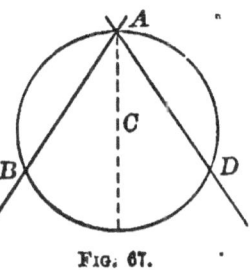

Fig. 67.

(c) If the centre be without the angle formed by the two chords, draw an auxiliary diameter and demonstrate.

The three cases are all the possible ones, and each having been demonstrated, the theorem is established.

58　ELEMENTS OF GEOMETRY.

Exercises.—1. Show that an inscribed right angle will be subtended by a semicircumference; an inscribed acute angle by an arc less than a semicircumference; and an inscribed obtuse angle by an arc greater than a semicircumference.

2. A secant separates a circle into two parts called *segments*. Show that all angles inscribed in a given segment are equal.

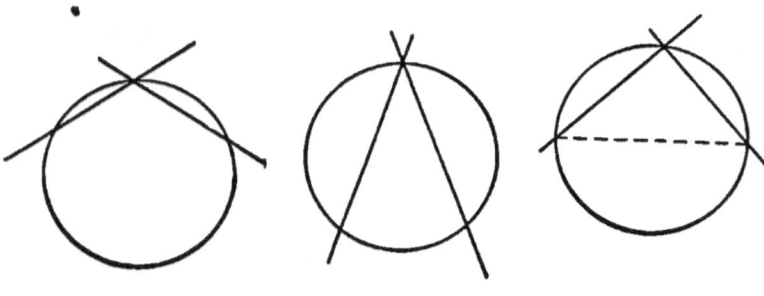

FIG. 68.

3. Show that angles inscribed in the two segments formed by a secant will be supplementary.

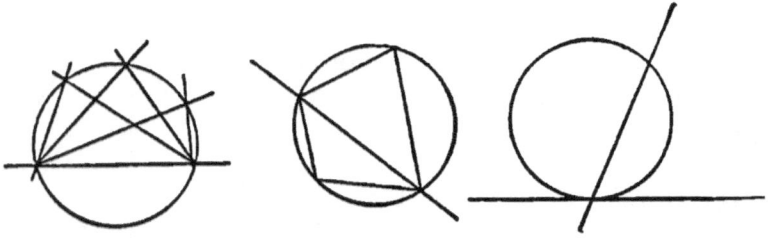

FIG. 69.

4. Show that the angles formed by a tangent, and a secant passing through the point of tangency, are measured by the halves of the intercepted arcs.

CIRCLES. 59

48. Theorem. *Parallel secants intercept equal arcs of the circumference of a circle.*

If $\overset{\frown}{AG}$ and $\overset{\frown}{BD}$ are equal, they may be brought to coincidence by rotation about a diameter perpendicular to the parallel chords.

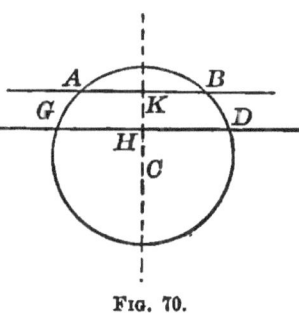

Fig. 70.

Acting upon this suggestion, draw a perpendicular through C. It will be a perpendicular bisector of both chords.

Revolve one semicircle upon the diameter as an axis. All the parts of the revolved figure will coincide with the parts of the figure that remains stationary. Hence $\overset{\frown}{AG} = \overset{\frown}{BD}$.

Exercises.—1. Demonstrate the theorem, making use of the principles established in § 45.

2. Establish it by the principles of § 47.

49. Theorem. *An angle formed by two secants which intersect within a circle, is measured by the half-sum of the arcs, subtended by the angle considered, and its vertical angle.*

Let AC and BD represent any two secants fulfilling the required conditions, and θ the angle considered.

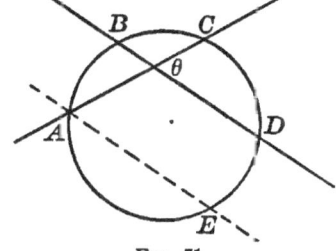

Fig. 71.

Analysis.—If a line be drawn parallel to BD, it will make with AC an angle θ; and if it be drawn so as to intersect the circumference, it will intercept equal arcs.

Demonstration. — Through A draw a parallel to BD.

$$\angle CAE = \theta,$$
$$\widehat{ED} = \widehat{AB},$$
$$\widehat{CE} = \widehat{CD} + \widehat{AB},$$
$$\theta = \angle CAE = \frac{\widehat{CE}}{2} = \frac{\widehat{CD} + \widehat{AB}}{2}.$$

Q. E. D.

Exercise. — Show the truth of the theorem by drawing auxiliary lines through the centre parallel to the secants.

50. Let AC and BD intersect within the circle. The $\angle \theta$ is measured by the half-sum of the arcs AB and CD.

The directions indicated by the arrowheads are positive.

If the secant BD be moved parallel to its initial position, so that CD be increased, AB will be diminished by a like amount.

When the position AD' shall have been reached, the $\angle \theta$ will be measured by the half of CD'.

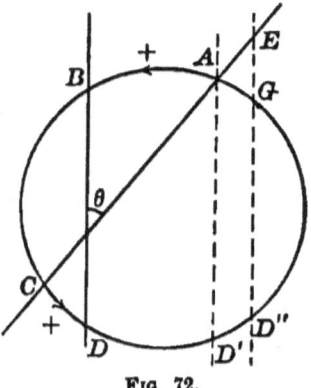

Fig. 72.

If a further movement be made, under the same conditions, and the position ED'' be reached, the $\angle \theta$ will not have changed, the arc measured from C will have increased, but the arc AG will be measured in a reverse sense from the arc AB, and is negative. In *length* it is equal to the arc $D'D''$, but under the circumstances is negative.

Hence the angle CED'' is measured by the half-sum of the intercepted arcs, — the arc that is *convex* toward the point E, being *negative*, and the one *concave* toward E being positive.

If the secant be further moved until it becomes a tangent, it is readily seen, by drawing AK parallel to the tangent line, that the $\angle CFT$ is measured by half the aggregate of the two arcs CT and AT; the arc AT being negative.

If the secant FC should now move parallel to itself until it should become a tangent, the angle which has remained the same would be measured by the aggregate of the arcs RMT and RQT; the latter being negative.

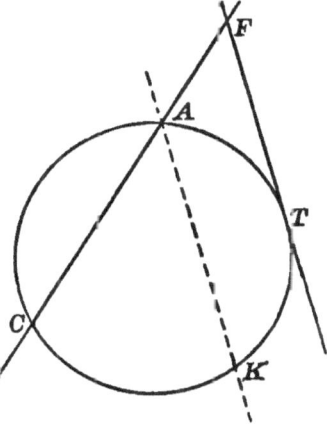

Fig. 73.

The student should show this by drawing an auxiliary line through one point of tangency parallel to the other tangent, or in any other way that he may choose.

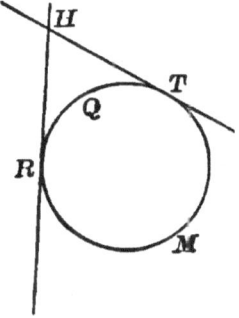

Fig. 74.

51. Theorem. *In the same or in equal circles, equal chords subtend equal arcs.*

Analysis. — *If* the equal chords do subtend equal arcs, the angles formed by joining their extremities with the centre will be equal, because angles at the centre are measured by the intercepted arcs.

Proof. — Draw the auxiliary lines indicated.

The △ ABC and DEC are equal, having the three sides of one equal to the three sides of the other.

$\therefore \angle ACB = \angle DCE$,

and because the angles are equal, the arcs subtending them will be equal.

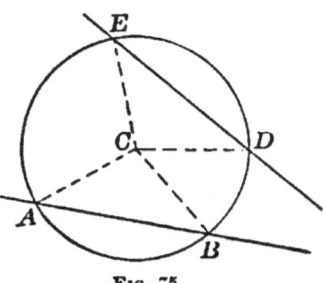

Fig. 75.

Q. E. D.

Exercise.—1. It is said that equal chords are equally distant from the centre. Is it true?

2. Demonstrate the converse of the theorem.

3. Demonstrate the *opposite* of the theorem.

52. Theorem. *In a circle the greater of two chords subtends the greater arc.*

Analysis.—If the greater chord does subtend the greater arc, when the arcs are brought so as to have two extremities coincident, the point D will lie beyond B from A.

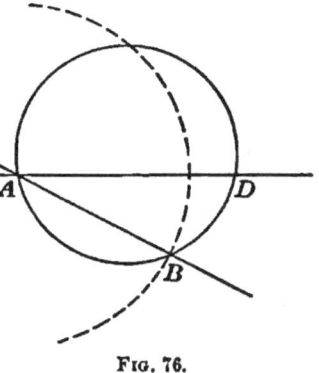

Fig. 76.

Demonstration.—The point D does lie beyond B; for if we construct an auxiliary circle having A for its centre and AB for its radius, the point D will fall outside of the auxiliary circle, for by hypothesis the distance AD is greater than the radius of the auxiliary circle.

Then the arc which is subtended by the greater chord will lie upon the arc subtended by the lesser chord throughout its entire length and extend beyond it. Hence it is greater, and the theorem is established.

53. Theorem. *The lesser of two unequal chords in a circle will be at the greater distance from the centre.*

The distances of the chords from the centre are the lengths of the perpendiculars from the centre to the chords.

If the chords be made to have an extremity of each coincident, the chords themselves will not coincide, but will occupy the relative positions indicated in the figure. AD, being the greater chord, subtends the greater arc, and so lies in such a position that *any* line drawn from C to any point in $\overline{AB}$ will cross $\overline{AD}$. Therefore the $\perp CH$ will cross AD at some point, as P, which does not coincide with K.

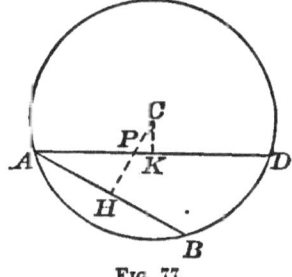

Fig. 77.

$$\therefore CH > CP > CK$$

(read: CH is greater than CP, which is greater than CK). Q. E. D.

Exercises.—1. All chords of equal length in a given circle are equally distant from the centre.

2. Find the maximum and minimum chords that may be drawn through a given point in a circle.

3. Establish the converse of the theorem.

4. Find the locus of any fixed point on a chord of given length.

54. Theorem. *If two circumferences intersect, the line of centres will be the perpendicular bisector of their common chord.*

Analysis.—If the line OO' be the perpendicular bisector of the chord CH, it must contain at *least* two points which are equally distant from C and H. Does it?

Demonstration.—O is equally distant from C and H; and O' is equally distant from C and H. Hence the theorem.

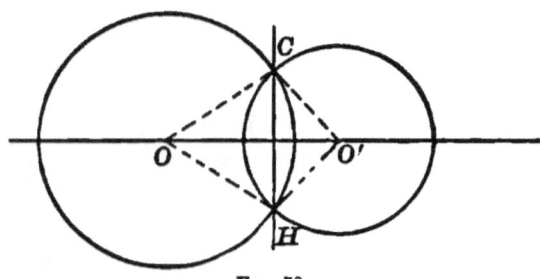

Fig. 78.

Exercises.—1. Show that if two circles intersect, the distance between their centres is less than the sum of their radii.

2. Show that if two circles are tangent to each other, they will have a common tangent line.

3. Show that the line of centres will pass through the point of tangency, and will equal the sum of the radii.

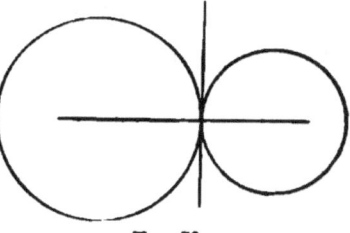

Fig. 79.

4. Show that if two circles are exterior the one to the other, the distance between centres will be greater than the sum of their radii.

55. Definitions. A *triangle* is inscribed within a circle when its vertices are on the circumference and its sides

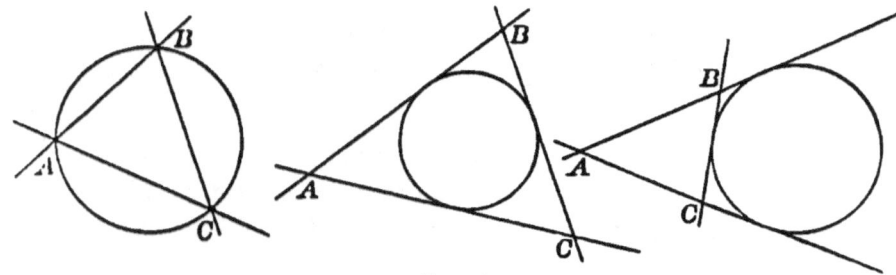

Fig. 80.

are chords. The *circle* is said to be **circumscribed** about the triangle.

A *circle* is **inscribed** within a triangle when the sides of the triangle are tangent to the circumference and the circle lies *within* the triangle.

A *circle* is **escribed** to a triangle when the lines forming the triangle are tangent to the circumference, but the circle does *not* lie within the triangle.

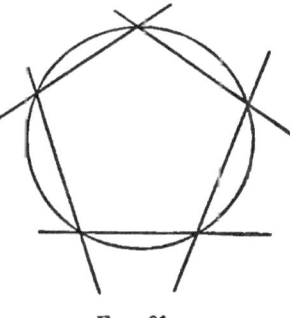

Fig. 81.

An *inscribed* polygon is a polygon the vertices of which lie on the circumference, and the sides of which are chords.

GENERAL EXERCISES.

1. PROBLEM. *Through a point without a circle to draw a tangent to the circumference.*

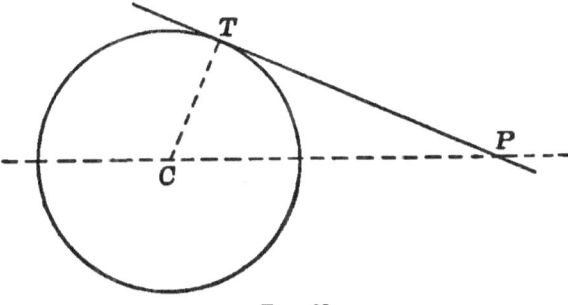

Fig. 82.

Analysis.— If PT were the required tangent through P, it would be perpendicular to a radius drawn to the point of tangency. *If* then we had CT drawn, $\angle CTP$ would be 90°. *If* we had a line CP, joining the two fixed points, a right triangle would be

F

formed having *CP* for its hypothenuse ; and *if* a circle *were* constructed with *CP* as its diameter, the circumference would pass through *T*.

Remark. — By the *analysis* a *sufficient* number of *necessary* and *previously established relations* have been determined to enable us *now* to make the *construction* which in its order will be the *reverse* of the *analysis*.

Construction. — Join the centre of the given circle to the given point. On this segment as a diameter construct the circumfer-

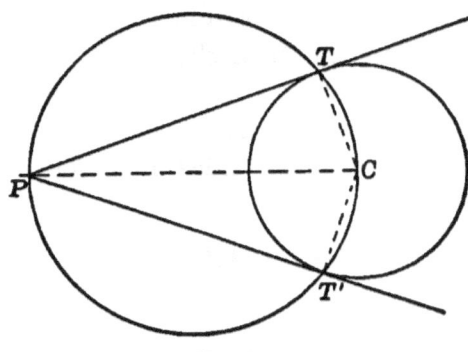

Fig. 63.

ence of a circle. It intersects the given circumference at *T* and *T'*. Draw *PT* and *PT'*. Both will be tangents, because they will be perpendicular to radii drawn from *C* to *T* and *T'*. $\angle PTC$ and $\angle PT'C$ are each inscribed in semicircumferences. Hence we have two constructions.

Discussion. — It is to be observed that if the point *P* should move to a greater distance from the given circle, the tangents would approach parallelism.

If the point *P* should move to a position on the circumference, the two tangents would coincide and form one.

If the point *P* were within the circle, the construction would not be possible.

CIRCLES.

2. Show that of all the points on the circumference of a circle, the nearest and the furthest from any given point will be on the line joining the given point with the centre of the given circle.

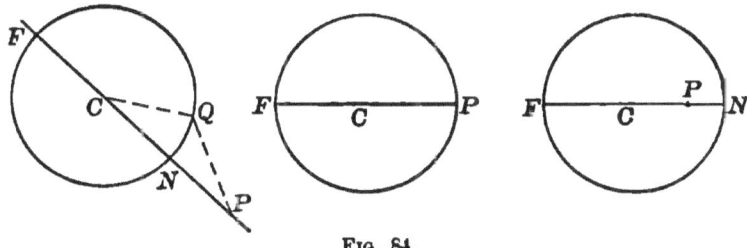

FIG. 84.

3. Show that of all the points on the circumference of a circle, the nearest and the furthest from another circumference will be on the line joining their centres.

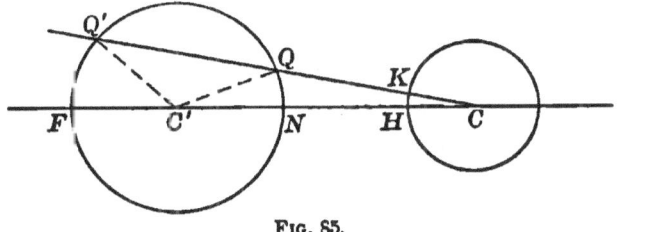

FIG. 85.

4. Construct a circle of given radius that *shall be tangent to two given circles*.

NOTE. — Make as many constructions as possible, and discuss the limitations of each.

6. Show that if from a point without a circumference tangents be drawn, the line joining the point with the centre bisects the angle formed by the tangents; bisects the angle formed by the radii drawn to the points of tangency; and bisects the arcs. The segments of the tangents are equal.

7. Inscribe a circle within a given triangle.

8. Use a triangle inscribed within a given circle to show that in any triangle having unequal sides the greater of two sides will lie opposite the greater angle.

9. Show that in a right triangle the difference between the sum of the perpendicular sides and the hypothenuse will be the diameter of the inscribed circle.

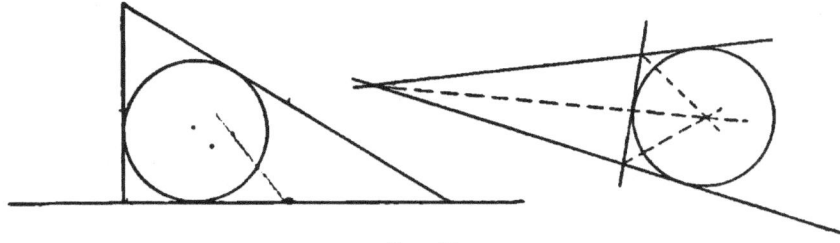

Fig. 86.

10. Show that the bisector of an interior angle and the bisectors of the non-adjacent exterior angles will pass through one point, and that point will be the centre of an escribed circle.

11. Show that if a regular hexagon be inscribed within a circle, each side will equal the radius of the circle.

12. Show that if any quadrangle be inscribed within a circle, the opposite angles will be supplementary.

13. Prove the converse.

14. Construct a segment of a given circle that shall be capable of containing a given angle.

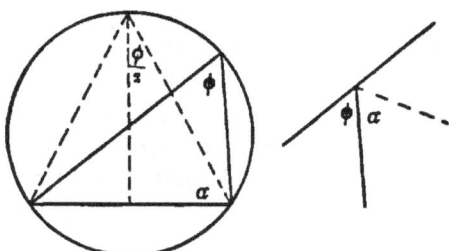

Fig. 87.

15. With a given line as a chord construct a segment of a circle that shall contain a given angle.

CHAPTER V.

56. Definitions. If a point move along a line, as AB, from any point, as A, toward B, it will in its course occupy the position of *every* point of the line as far as we may conceive it as moving.

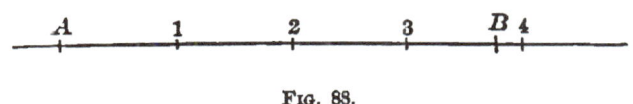

FIG. 88.

If we fix our attention upon B, and speak of the distance of B from A, it is a *definite* thing, and is perfectly understood.

For purposes of description and comparison we frequently take some convenient *unit* of measure, and apply it to the distance. If the distance be a day's journey, we use the *mile* or the *kilometre*. If it be a distance, as in the figure, between points on the page of this book, we use *inches* or *centimetres*.

Remembering that the point in moving along the line from A to B occupies an infinite number of positions, one sees that the chances that the extremity of the measuring unit will not fall at B are as infinity (∞) to one.

In the above figure, we would, "roughly speaking," say that B was 3 centimetres from A. If we desired, for any reason, more accurately to describe the distance, we should descend to fractional parts of this unit; the fractional part being less than the distance by which we failed to reach B when using the entire unit. Again, the chances are, as infinity to one, that the new unit of measure will *not* fall on B. The subdividing of the unit may be carried to any extent, depending entirely upon the required accuracy of description.

The most convenient subdivision is the decimal one.

If the applied unit or any of its subdivisions have their extremities at B, the distance AB and the unit are said to be *commensurable*. In general, however, if a point, as B, is taken at random on the line AB, its distance from A will not be *commensurable* in terms of any established unit. The distance is then said to be *incommensurable* with the unit.

It might be commensurable with respect to one unit and incommensurable with respect to another unit.*

A point may be so assumed that its distance from A shall be commensurable in terms of any unit that may be selected.

---

* NOTE IN ILLUSTRATION. — If we undertake to express *decimally* the distance of a point from A that is distant therefrom $\frac{2}{3}$ of the unit distance, we can only approximate to it. The first approximation would be .6, a nearer one would be .66, a still nearer .666. We might continue annexing decimal places as long as we please and we should never, in that way, reach the point that is $\frac{2}{3}$ of a unit's distance from A; although at each step we should come nearer the point.

$\frac{2}{3}$ is said to be the LIMIT toward which we approach as we increase the number of decimal places in .6666 . . . .

AREA. 71

57. We know from § 19 that parallel lines are everywhere equally distant from each other. Let AB and CD in the figure represent two parallel lines, and AC a line perpendicular to AB and CD. Represent the segment AC by (h).

If we cause the line AC to move parallel to itself a distance (b), the extremities of (h) remaining in AB and CD, the surface swept over by the segment (h) is described as "the area bh."

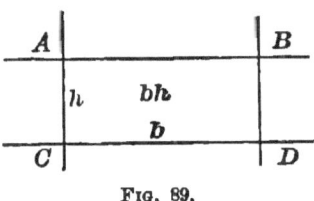

Fig. 89.

AC and BD are parallel, and if the line CD, perpendicular to them and remaining always parallel to its initial position, should move to the position AB, the segment (b) would sweep over the area (hb). The areas swept over being the same, we have $bh = hb$.

Either side of the rectangle $ACDB$ may be called the base; ordinarily it would be the lower one in the figure. The perpendicular distance between the base and its parallel is called the altitude. Sometimes this parallel is called the upper base.

The rectangle is accurately described by bh or hb.

In general, b and h are incommensurable with any assumed unit of length, and the area bh is incommensurable with the square, having that unit for its side. But for convenience of numerical description or comparison some unit square is taken and applied to the rectangle. If it be large tracts of land that are being considered, we use the acre or the hectare. If it be the areas of rectangles on a page of this book, the square inch would be appropriate.

If we should assume *CM* as the side of a unit square, and should lay it off from *C* towards *A* as many times as possible, an extremity would fall within a unit's length of *A*, as at *H*. If we lay off the same unit of length from *C* toward *D*, as many times as possible, an extremity would fall within a unit's length of *D*, as at *K*.

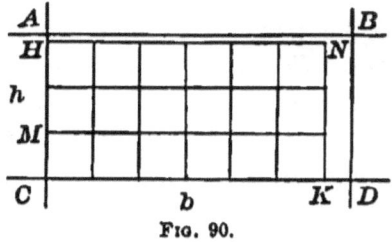

Fig. 90.

If the line *CH* move parallel to itself and so that each point moves on a perpendicular to *CA*, until it occupy the position *KN*, it will have swept over an area expressed by $CK \times KN$. It will be commensurable with the assumed unit area, and the number of times it will contain that unit is expressed by the product of the number of units of length in *CH* by the number of units of length in *CK*.

If a nearer approximation is desired, the former unit of area is subdivided, so that a side of the new comparison square will be less than the distance by which in the preceding instance we failed of reaching either *AB* or *BD*.

We shall thus have an increased commensurable area, which will be a nearer approximation to the area *bh*.

This subdivision of the unit after the manner above indicated may be carried as far as we please, and a commensurable area be expressed in the terms of some measuring unit, which shall approximate as nearly as we may please to the incommensurable area *bh*.

bh is not necessarily incommensurable, but is generally so.

AREA.

58. Let A represent the area of the rectangle $CDEF$, and A' the area of the rectangle $HJKL$. Then $A = bh$, and $A' = b'h'$.

Dividing the members of one equation by the members of the other, we have, $\dfrac{A}{A'} = \dfrac{bh}{b'h'}$;

i.e. two rectangles have the same ratio as the products of their bases and altitudes.

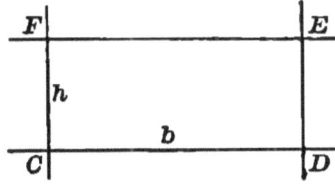 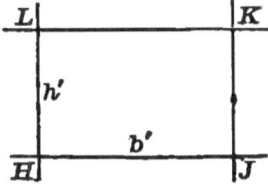

Fig. 91.

If h and h' happen to be equal,
$$\frac{A}{A'} = \frac{b}{b'};$$

i.e. when the altitudes are equal, the ratio of the areas is the same as the ratio of the bases.

If instead of h and h' being equal, it happen that b and b' were equal, we would have,
$$\frac{A}{A'} = \frac{h}{h'};$$

i.e. the ratio of the areas, when the bases are equal, would be the same as the ratio of the altitudes.

59. Let $ABCD$ represent any parallelogram. If through a pair of adjacent vertices, as A and D, lines be drawn perpendicular to the line AD, a rectangle $AKHD$ will be formed. The area $ABHD$ is common to the rectangle $AKHD$ and the parallelogram $ABCD$. The parts of each not common to the other are the triangles AKB and DHC. These triangles are congruent, therefore the surface $AKHD$ equals the surface $ABCD$.

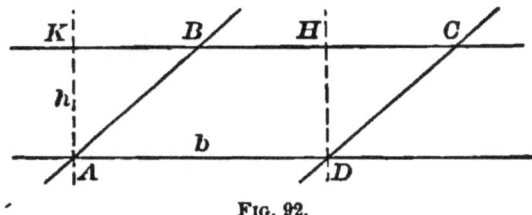

Fig. 92.

The surface $AKHD = bh$. Therefore the surface $ABCD = bh$, in which b is one side of the parallelogram, and h is the segment of the perpendicular to (b) that is included between (b) and the line forming the opposite side of the parallelogram. Therefore,

The surface of a parallelogram will be represented by bh.

Exercises.—1. Show that the ratio of the surfaces of two parallelograms to each other equals the ratio of the products of bases and altitudes.

2. If the altitudes are equal, the surfaces will have the same ratio as the bases.

3. Show that if lines be drawn through two vertices of a triangle parallel to the opposite sides, a parallelogram will be formed, the surface of which will be double that of the triangle.

60. By the last exercise in the preceding article it is shown that a triangle will be half of a certain parallelogram. By § 59, any parallelogram is equivalent to a rectangle having the same base and altitude, and the area of a rectangle is represented by (bh), in which (b) represents the base, and (h) represents the altitude. Therefore,

The surface of a triangle will be represented by $\dfrac{bh}{2}$.

Exercises. — 1. Show that the ratio of the surfaces of any two triangles to each other equals the ratio of the products of their bases and altitudes.

2. Show that if their bases are equal, the surfaces will be to each other as their altitudes.

3. Show that if their altitudes are equal, the surfaces will be to each other as their bases.

4. Show that if two triangles have their bases in the same line and their vertices at the same point, their areas are to each other as their bases.

5. Show that if two triangles have their bases in the same line and their vertices in a parallel line, the ratio of their areas equals the ratio of their bases.

61. Theorem. *The area of a trapezoid equals the half-sum of its parallel sides multiplied by the perpendicular.*

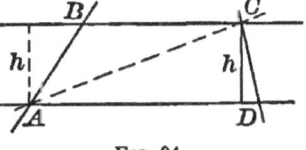

Fig. 94.

Let $ABCD$ represent the trapezoid. Draw a diagonal as AC. Each triangle composing the trapezoid will have the same altitude (h); hence,

$$\text{Area } ABC = \tfrac{1}{2} BC \cdot h.$$

$$\text{Area } ACD = \tfrac{1}{2} AD \cdot h.$$

$$ABC + ACD = \tfrac{1}{2}(BC) \cdot h + \tfrac{1}{2}(AD) \cdot h.$$

But $\quad ABC + ACD = $ area of the trapezoid.

$$\therefore \text{Area } ABCD = \tfrac{1}{2}(BC + AD) \cdot h. \quad \text{Q. E. D.}$$

Exercise. — Show that the segment of the line joining the middle points of the non-parallel sides of a trapezoid will equal the half-sum of the parallel sides.

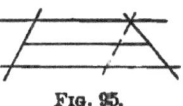

Fig. 95.

62. Theorem. *If a line be drawn parallel to any side of a triangle, the other sides will be separated into segments, which will form equal ratios.*

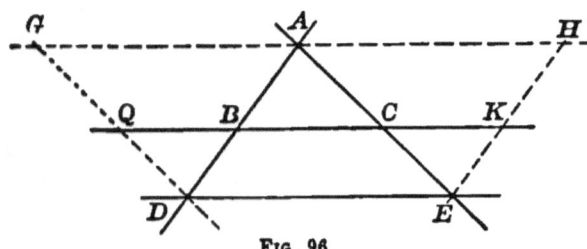

Fig. 96.

Let ADE represent the given triangle, and BC a line drawn parallel to DE. Draw the auxiliary lines through the vertices parallel to the opposite sides.

By Exercise 2, § 59,

$$\frac{\Box ABKH}{\Box ADEH} = \frac{AB}{AD}, \qquad (1)$$

$$\frac{\Box ACQG}{\Box AEDG} = \frac{AC}{AE}. \qquad (2)$$

But $\Box ABKH = \Box ACQB$, since their bases BK and QC are each equal to DE, and their altitudes are the same. Also $\Box ADEH = \Box AEDG$, since they have the same base and equal altitudes.

The first members of equations (1) and (2) being equal, the second members are, and

$$\frac{AB}{AD} = \frac{AC}{AE}. \qquad \text{Q. E. D.}$$

Exercises.—1. Show that $\dfrac{AB}{BD} = \dfrac{AC}{CE}$.

2. Show that $\dfrac{AB}{AC} = \dfrac{AD}{AE}$.

3. Show that $\dfrac{AB}{AC} = \dfrac{BD}{CE}$.

PROPORTIONAL DIVISION. 77

4. Show that if a line be drawn through the middle point of one side of a triangle, parallel to a second side, it will bisect the third side.

5. Show that the new triangle formed in Ex. 4 will be one-fourth the area of the original triangle.

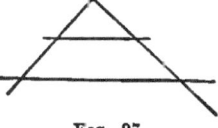

FIG. 97.

63. PROBLEM. Establish the converse of the theorem in § 62, viz.: If a straight line be drawn through points that separate two sides of a triangle into proportional* segments, it will be parallel to the third side.

Recalling the fact that the *reductio ad absurdum* is a particularly appropriate method for determining the truth or falsity of the converse of a theorem, we proceed as follows:

If the line BC be the line through the points of proportional division, and *if* it be *not* parallel to DE, let us draw a line BQ through B that shall be parallel to DE. By § 62,

$$\frac{AB}{AD} = \frac{AQ}{AE}. \qquad (1)$$

But $\dfrac{AB}{AD} = \dfrac{AC}{AE}$ by hypothesis. Multiplying each member of both equations by AE, we have, from (1),

$$\frac{AB}{AD} \cdot AE = AQ,$$

from (2), $\qquad \dfrac{AB}{AD} \cdot AE = AC.$

---

* NOTE. — Four quantities which may form two equal ratios are proportional. Frequently the numerators of the fractions used in expressing a proportion are called *antecedents,* and the denominators *consequents.*

78 ELEMENTS OF GEOMETRY.

The first members are the same, hence the second must be, and
$$AQ = AC.$$

The supposition that the line BC is not parallel to DE thus leads us to an absurdity, and the supposition that BC is not parallel to DE is an erroneous one.

64. Theorem. *If two triangles have the three angles of one equal to the three angles of the other, each to each, the ratio of any two sides of one will equal the ratio of the corresponding* sides of the other.*

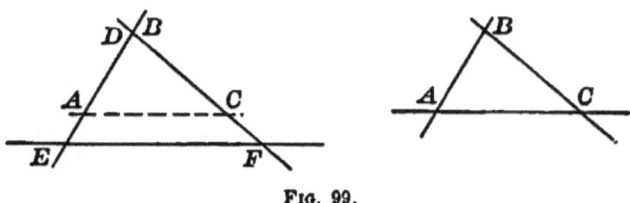

Fig. 99.

Let ABC and DEF represent two mutually equiangular triangles.

Let
$$\angle ABC = \angle EDF,$$
$$\angle BAC = \angle DEF,$$
$$\angle BCA = \angle DFE.$$

Superimpose the $\triangle ABC$ upon the $\triangle EDF$, so that the $\angle ABC$ shall coincide with the $\angle EDF$. The interior angles at A and at E are equal by hypothesis. Then AC is parallel to EF.

$$\therefore \frac{BA}{DE} = \frac{BC}{DF}. \qquad (\S\ 62),\ (1)$$

---

* Note. — In mutually equiangular triangles, the sides opposite equal angles are called " corresponding."

If the triangles had been superimposed so that the interior angles at C and F had been made coincident, AB would have been parallel to ED.

$$\therefore \frac{CA}{FE} = \frac{CB}{FD}. \qquad (\S\ 62),\ (2)$$

Either by superposition or by the equality axiom, using equations (1) and (2), we may show that

$$\frac{BA}{DE} = \frac{CA}{FE}. \qquad \text{Q. E. D.}$$

65. Definitions. Similar figures are those in which the angles of one are equal to the angles of the other, and the corresponding sides are proportional.

In a figure other than a triangle, corresponding sides are best described as being those that lie between mutually equal angles.

Exercises. — **1.** Show that if two triangles have the three sides of one perpendicular respectively to the three sides of another, the two triangles will be mutually equiangular, and hence similar.

2. Show that if two triangles have an angle in each equal, and an angle in one the supplement of an angle in the other, the ratio of the sides opposite the equal angles is equal to the ratio of the sides opposite the supplementary angles.

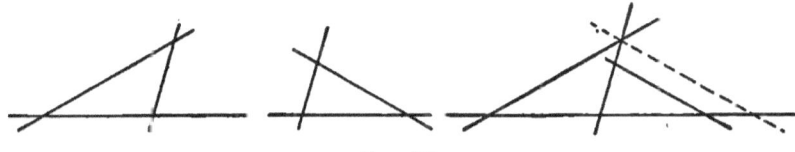

Fig. 100.

3. Show that if two triangles have the three sides of one parallel to the three sides of the other, the triangles will be equiangular and therefore similar.

66. THEOREM. *If two triangles have an angle in each equal and the including sides proportional, they are similar.*

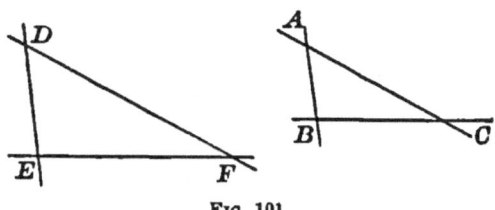

FIG. 101.

Let the interior angles at A and D be equal, and $\dfrac{AB}{DE} = \dfrac{AC}{DF}$. If ABC be superimposed upon DEF so that the interior angle at A shall coincide with the interior angle at D, we shall have B and C fall at points on DE and DF so as to divide them proportionally.

Then by § 63, BC will be parallel to EF, and

$\angle ABC = \angle DEF$
$\angle ACB = \angle DFE$ and $\dfrac{AB}{DE} = \dfrac{BC}{EF} = \dfrac{AC}{DF}$. Q. E. D.

67. THEOREM. *If the three sides of a triangle are proportional to the three sides of another triangle, they will be mutually equiangular, and hence similar.*

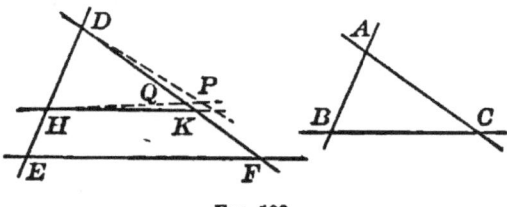

FIG. 102.

This is the converse of § 64.

By hypothesis, $\dfrac{AB}{DE} = \dfrac{AC}{DF} = \dfrac{BC}{EF}$.

If the angles are not equal, when we attempt superposition, causing the side AB to fall on the line DE, as at DH, the vertex C will fall at some point as P, not on DF.

Through the point H, draw HK parallel to EF, and writing DH for its equal AB, we have:

By § 62, $\qquad \dfrac{DH}{DE} = \dfrac{DK}{DF}.$

But by hypothesis, $\dfrac{DH}{DE} = \dfrac{AC}{DF}.$

$$\therefore DK = AC = DP.$$

Also by § 62, $\qquad \dfrac{DH}{DE} = \dfrac{HK}{EF}.$

But by hypothesis, $\dfrac{DH}{DE} = \dfrac{BC}{EF}.$

$$\therefore HK = BC = HP.$$

Hence the △ DHP and DHK having three sides of one equal to the three sides of the other are equal in all their parts, and $\angle HDP = \angle HDK$. Therefore the supposition that the vertex C did not fall on DF is an erroneous one. It does fall on DF, and § 66 applies to establish the similarity. Q. E. D.

68. Constructions. 1. *To divide a given segment of a straight line into a given number of equal parts.*

Let AB represent the segment that is to be separated into equal parts, say four.

Draw any line AK in a convenient position. With the dividers, or any convenient measure, lay off a convenient distance AJ. Lay off $JT = TL = LK = AJ$. Join KB. Through L, T, and J draw parallels to KB.

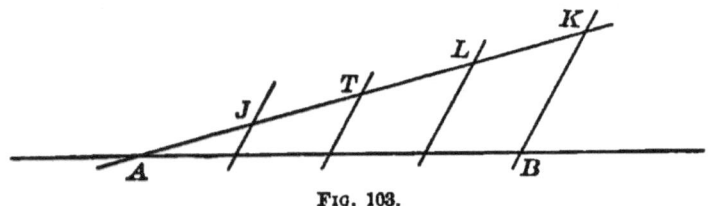

Fig. 103.

By § 62, AB will be separated into four equal parts.

Q. E. F.

2. *To divide a given segment of a straight line into parts that shall be proportional to any given segments.*

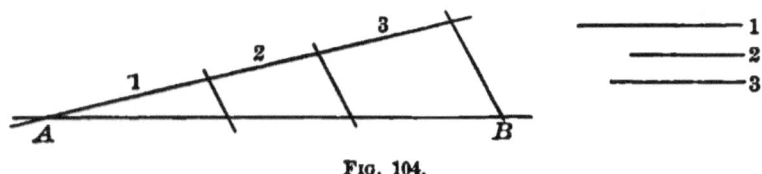

Fig. 104.

3. *To draw a triangle that shall have a given perimeter and shall be similar to a given triangle.*

4. *To find a fourth proportional to three given segments of straight lines.*

If x represent the line to be determined,

$$\frac{a}{b} = \frac{c}{x}$$

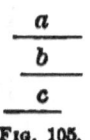

Fig. 105.

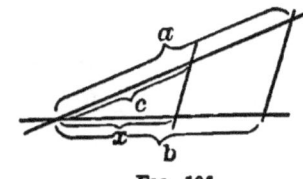

Fig. 106.

MEDIANS. 83

The form immediately suggests an application of Exercise 2, § 62.

5. *To construct x in* $x = \dfrac{b^2}{a}$.

$$\frac{x}{b} = \frac{b}{a}, \text{ or } \frac{a}{b} = \frac{b}{x}.$$

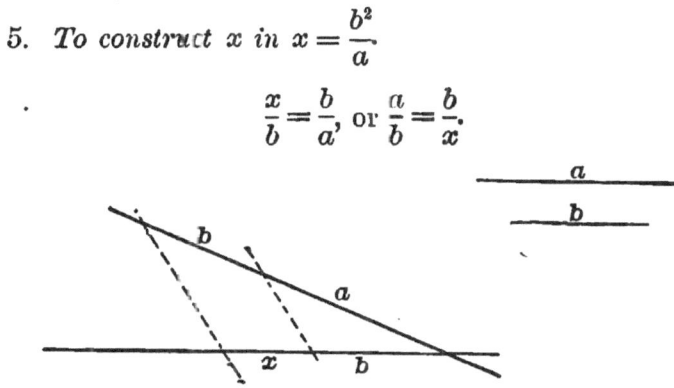

FIG. 107.

69. Definition. If a secant of a triangle pass through a vertex and the middle of the sides opposite, the segment between these points is called a **median**. There will be three medians in every triangle.

Problems. — 1. Show that two medians of a triangle trisect each other; *i.e.* separate each other into two segments, one of which is one-third the whole median.

The analysis of the problem suggests the following:

Draw PQ, joining the middle points of AI and BI. It will be parallel to BA and equal to $\dfrac{BA}{2}$. Draw DE; it will be parallel to BA and equal to $\dfrac{BA}{2}$.

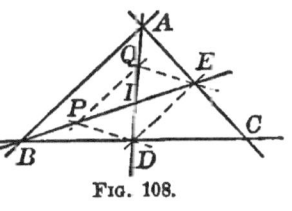

FIG. 108.

Therefore PQ and DE are parallel and equal, and $PQED$ is a parallelogram (Ex. 2, § 38).

Its diagonals bisect each other, or $QI = ID$. But $QI = AQ$ by construction; therefore $AQ = QI = ID$.

For the same reasons $BP = PI = IE$.

Q. E. D.

84   ELEMENTS OF GEOMETRY.

2. Show that the third median would also pass through I.

3. Having given the three medians of a triangle, to construct it.

Analysis. — If ABC were the required triangle, with m', m'', and m''' the given medians; and if we should draw QR, it would be parallel to BC and equal to $\frac{BC}{2}$. If RH be taken equal to QR, and the point H be joined with A, P, and C, three parallelograms will be formed, from which we may establish the fact that APH will be a triangle, having the three medians for its sides.

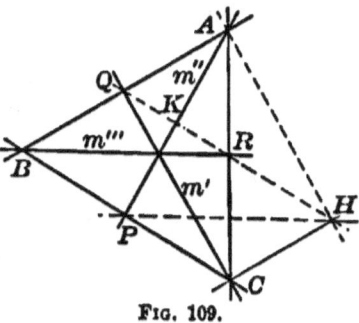

Fig. 109.

K will be the middle point of AP.

Construction. — Form a triangle with the three medians as sides. Draw a median of this triangle through any vertex. Pro-

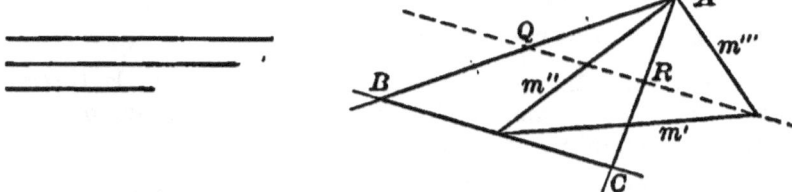

Fig. 110.

duce it one-third of its length. Through either of the other vertices, as A, draw AQ and AR, and lay off $QB = AQ$ and $RC = AR$. Join BC. ABC will be the required triangle.

4. If through any point three lines be drawn intersecting parallels, the segments will be proportional.

5. Establish the converse of Problem 4.

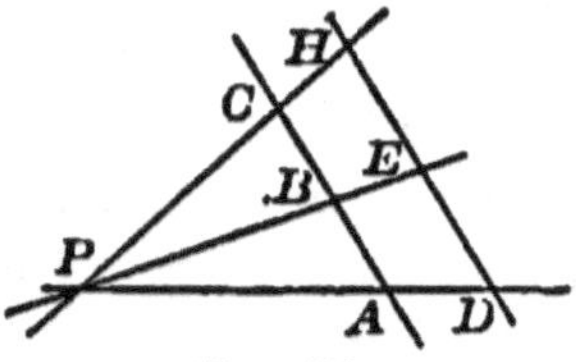

Fig. 111.

6. How does the bisector of an interior angle of a triangle divide the opposite side?

MEDIANS.

If AD bisects the interior angle at A, we will have two triangles, BAD and CAD having an angle in each equal, and other two angles supplementary.

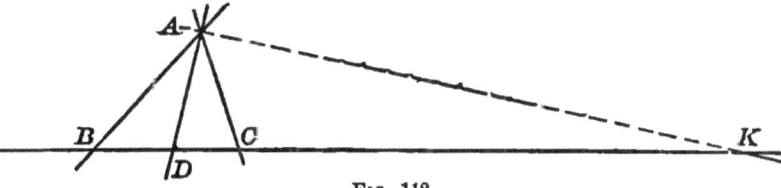

Fig. 112.

By Ex. 2. § 65, we have $\dfrac{BD}{DC} = \dfrac{AB}{AC}$.

Hence the opposite side is divided into segments proportional to the adjacent sides.

7. How does the bisector of an exterior angle of a triangle divide the opposite side?

If AK bisects the exterior angle at A, we have two triangles, BAK and CAK having the $\angle K$ in common and the $\angle CAK$ supplementary to the $\angle BAK$.

By Ex. 2, § 65, $\dfrac{BK}{CK} = \dfrac{AB}{AC}$.

Hence the two segments formed by the point of intersection and the other vertices will be proportional to the sides having their vertex at the vertex of the bisected angle.

8. Show that $\dfrac{BD}{DC} = \dfrac{BK}{CK}$.

GENERAL EXERCISES.

1. On a given segment as one side construct a parallelogram similar to a given parallelogram.

2. Show that similar polygons may be separated by auxiliary lines into similar triangles.

3. Show that circles are similar figures.

4. Show that the corresponding altitudes of similar triangles will be proportional to any set of corresponding sides.

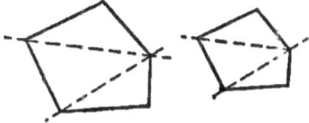

Fig. 113.

5. Show that the radii of circles inscribed in similar triangles are proportional to the corresponding sides.

6. Show that the same relation exists between the diameters of circumscribed circles.

7. Show that the same relation exists between the radii of the corresponding escribed circles.

8. Show that the corresponding altitudes are to each other as the corresponding medians.

9. Show that the corresponding angle bisectors are to each other as the perimeters of the triangles.

NOTE. — There are three principles in the elements of geometry that are more prominent than any others.

We have now established the first of these, as follows:

Corresponding lines of similar figures are to each other as ANY OTHER *corresponding lines.*

The second of these great principles, which will be established in the next chapter, is:

Similar areas are to each other as the squares of any corresponding lines.

The third, which will be established in Chapter XIII., is:

Similar volumes are to each other as the cubes of any corresponding lines.

CHAPTER VI.

70. Theorem. *The square constructed on the sum of two segments of a line equals the sum of the squares on the two segments* PLUS *twice the rectangle of the two segments.*

Place the two segments so that MN shall be their sum.

On MN as one side, construct the square MH.

At P erect the $\perp PG$. On MK lay off $MD = a$. Draw DE ∥ to MN. The student will show that:

$$MDIP = a^2,$$
$$IGHE = b^2,$$
$$DKGI = ab,$$
$$IENP = ab.$$

Adding, we have $(a + b)^2 = a^2 + b^2 + 2ab$; a relation that we are already familiar with in algebra. Q. E. D.

71. Theorem. *The square constructed on the difference of two segments of a line equals the sum of the squares on the two segments* MINUS *twice the rectangle of the two segments.*

Place the two segments so that MN shall be their difference, MP representing one segment a, and PN representing the other segment b.

87

On MN construct the square MI; it will be the square on $a-b$.

On a construct the square MH. Lay off DJ and EG each equal to b. Join JG.

$$JG = b^2,$$
$$JK = ab,$$
$$NH = ab.$$

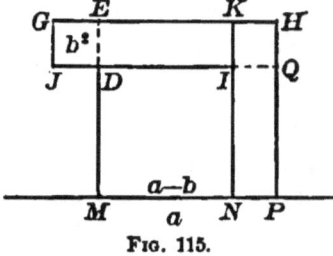

Fig. 115.

From the figure we see that $(a-b)^2 = a^2 + b^2 - 2ab$.

Q. E. D.

72. Theorem. *The rectangle having the sum of two segments for one side and the difference of the same segments for an adjacent side, equals the difference of the squares on the two segments.*

With CE and CN (as the sum and difference respectively of the two segments) for adjacent sides construct the rectangle CG.

On a construct the square CJ, and on HJ the square HK.

The rect. $NK = (a-b)b$
$\qquad$ = the rect. DG.

rect. CH + rect. $NK = a^2 - b^2$,
$\qquad\qquad$ = rect. CH + rect. DG
$\qquad\qquad$ = rect. $CG = (a+b)(a-b)$.

$\therefore (a+b)(a-b) = a^2 - b^2$;

another form that we remember from algebra.

Q. E. D.

SQUARES ON SEGMENTS. 89

73. THEOREM. *If squares be constructed upon the three sides of a right triangle, the square on the hypothenuse equals the sum of the squares on the other two sides.*

Let ABC represent the given right triangle, right-angled at B. On the side AB construct the square $AEDB$ exterior to the triangle, and on the side BC construct the square $BCKG$, overlying a part of the triangle.

At the vertex of the angle A erect a perpendicular to AC. It will intersect ED in some point as at J.

At C erect a perpendicular to AC; and through J draw a parallel to AC, forming the rectangle $AJHC$.

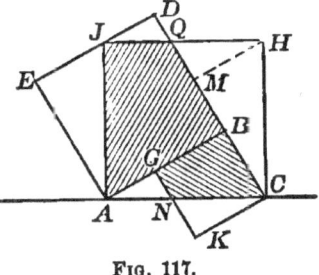

Fig. 117.

The $\triangle AEJ$ is similar to the $\triangle ABC$: the sides of one being perpendicular to the sides of the other. They are more than similar; the corresponding sides AE and AB, being sides of the same square, are equal. Hence the triangles are equal, and $AJ = AC$.

The rectangle $AJHC$ is therefore a square on the hypothenuse AC.

Draw *one* auxiliary line, viz. a $\perp HM$ from H to the line CD.

The shaded portion of the square on AB is also a part of the square on AC.

The shaded portion of the square on BC is also a part of the square on AC.

$\triangle AEJ = \triangle CMH$ (?), and may be placed so as to coincide with it.

△ CKN = △ HMQ (?), and may be placed so as to coincide with it.

△ JDQ = △ AGN (?), and may be placed so as to coincide with it.

Hence all the parts of the squares on AB and BC have been placed so as to coincide with the square on AC without repetition. And, furthermore, the square on AC has been completely covered by the parts of the other two squares. Hence the theorem.

Exercises. — 1. The side of a square is 1; what will the diagonal be?

2. The side of a square is a; what will the diagonal be?

3. Show that the square on either of the perpendicular sides of a right triangle equals the square on the hypothenuse, minus the square on the other perpendicular.

74. Definition. If in a plane, lines be drawn through the extremities of a segment, perpendicular to a given line, the intercepted portion of the given line is the perpendicular (orthogonal) projection of the given segment on the given line.

In the accompanying figure, CD is the orthogonal projection of AB.

If through A and B parallels be drawn obliquely to the line CD, the intercept EF is an *oblique* projection of AB.

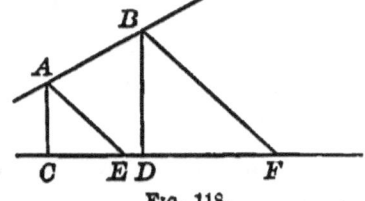

Fig. 118.

There can be but one perpendicular projection, but there may be an infinite number of oblique projections.

Unless otherwise specified, the perpendicular projection is the one meant when the word *projection* is used.

SQUARES ON THE SIDES OF TRIANGLES. 91

75. Theorem. *If a triangle be obtuse, the square constructed on the side opposite the obtuse angle equals the sum of the squares constructed on the other two sides, plus twice the rectangle formed by one of them and the projection of the other upon it.*

Let ABC represent the obtuse triangle.

Through B draw BP perpendicular to AC.

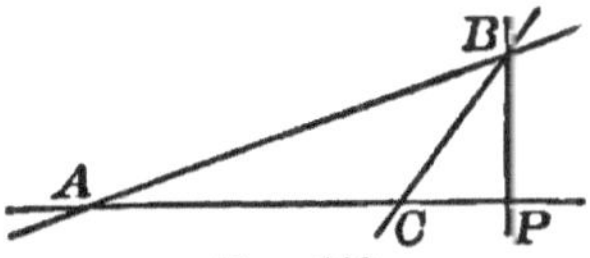

FIG. 119.

By § 73, $\quad \overline{AB}^2 = \overline{AP}^2 + \overline{PB}^2.$

By § 70, $\quad \overline{AP}^2 = (\overline{AC} + \overline{CP})^2$
$\qquad\qquad = \overline{AC}^2 + 2\,\overline{AC}\cdot\overline{CP} + \overline{CP}^2.$

By § 73, $\quad \overline{PB}^2 = \overline{CB}^2 - \overline{CP}^2.$

$\therefore \overline{AB}^2 = \overline{AB}^2 + \overline{PB}^2 = \overline{AC}^2 + \overline{CB}^2 + 2\,\overline{AC}\cdot\overline{CP}.$

Q. E. D.

Exercises. — 1. Prove the theorem by letting fall a perpendicular from A to the line BC.

2. Show that if a given point outside a circle be joined to a point in the circumference, M, and the point M be caused to move continuously about the circumference from the nearest point N, until it shall return to N, the length of the segment PM will vary continuously between the limits PN and PF.

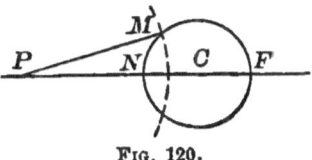

FIG. 120.

76. Theorem. *The square constructed on the side opposite an acute angle of* ANY *triangle equals the sum of the squares on the other two sides* MINUS *twice the rectangle formed by one of them and the projection of the other upon it.*

92   ELEMENTS OF GEOMETRY.

Let ABC in either figure represent an oblique triangle, the interior angle at C being acute.

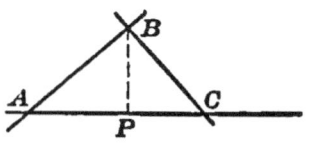 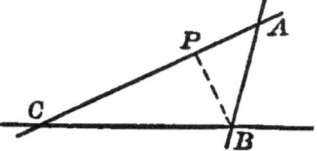

Fig. 121.

By § 73, $\overline{AB}^2 = \overline{AP}^2 + \overline{PB}^2.$

By § 71, $\overline{AP}^2 = \overline{AC}^2 + \overline{PC}^2 - 2\,\overline{AC}\cdot\overline{PC}.$

By Ex. 3, § 73, $\overline{PB}^2 = \overline{BC}^2 - \overline{PC}^2.$

∴ $\overline{AB}^2 = \overline{AP}^2 + \overline{PB}^2 = \overline{AC}^2 + \overline{BC}^2 - 2\,\overline{AC}\cdot\overline{PC}.$

Q. E. D.

NOTE. — The accompanying figure is a convenient one for helping the memory to retain the relations established in §§ 73, 75, 76.

Let ACH be a right triangle.
Let ACB be an obtuse triangle having AC and CB equal to AC and CH of the right triangle.
Let ACK be an acute triangle having AC and CK equal to AC and CH of the right triangle.

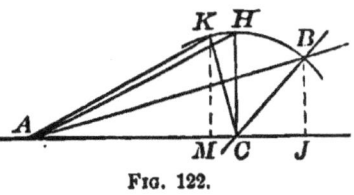

Fig. 122.

By either General Ex. 3, of Chapter II., or Ex. 2, § 75,

$$AB > AH > AK,$$
$$\overline{AH}^2 = \overline{AC}^2 + \overline{CH}^2,$$
$$\overline{AB}^2 = \overline{AC}^2\,\overline{BC}^2 + + 2\,\overline{AC}\cdot\overline{CJ},$$
$$\overline{AK}^2 = \overline{AC}^2 + \overline{CK}^2 - 2\,\overline{AC}\cdot\overline{MC}.$$

AREAS OF SIMILAR TRIANGLES. 93

77. Theorem. *If in a right triangle a line be drawn through the vertex of the right angle perpendicular to the hypothenuse, it will separate the triangle into two triangles similar to the given one, and hence similar to each other.*

The △ *AFB* and *ABC* have the interior angle at *A* in common, and ∠ *AFB* = ∠ *ABC*, each being equal to 90°. The third angles must be equal. Hence the triangles are similar (§ 64).

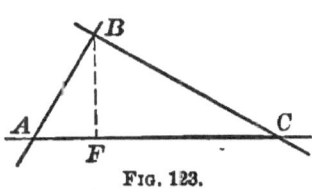

Fig. 123.

In the same way △ *CFB* and *CBA* are shown to be similar.

The △ *AFB* and *CFB* having the angles of each equal to the angles of *ABC*, will be mutually equiangular, and hence similar. Q. E. D.

(*a*) By reason of the similarity of the △ *AFB* and *CFB*, we have:

$$\frac{AF}{FB} = \frac{FB}{FC}.$$

$$\therefore \overline{AF} \cdot \overline{FC} = \overline{FB}^2, \tag{1}$$

or, *The rectangle of the segments of the hypothenuse equals the square of the perpendicular.*

(*b*) By reason of the similarity of the △ *AFB* and *ABC*, we have:

$$\frac{AF}{AB} = \frac{AB}{AC}.$$

$$\therefore \overline{AF} \cdot \overline{AC} = \overline{AB}^2. \tag{2}$$

Also from △ *CFB* and *ABC*, we have:

$$\frac{FC}{CB} = \frac{CB}{AC}.$$

$$\therefore \overline{FC} \cdot \overline{AC} = \overline{CB}^2, \tag{3}$$

or, *The square of either side about the right angle equals the rectangle of the whole hypothenuse and the adjacent segment.*

(c) Dividing the members of Eq. 2 by those of Eq. 3, we have:

$$\frac{\overline{AF}\cdot\overline{AC}}{\overline{FC}\cdot\overline{AC}} = \frac{\overline{AB}^2}{\overline{CB}^2}.$$

$$\therefore \frac{\overline{AF}}{\overline{FC}} = \frac{\overline{AB}^2}{\overline{CB}^2}, \tag{4}$$

or, *The ratio of the segments equals the ratio of the squares of the corresponding sides about the right angle.*

Adding the members of Eq. 2 to the members of Eq. 3, we have:

$$\overline{AF}\cdot\overline{AC} + \overline{FC}\cdot\overline{AC} = \overline{AB}^2 + \overline{CB}^2.$$

$$(\overline{AF}+\overline{FC})\,\overline{AC} = \overline{AB}^2 + \overline{CB}^2.$$

$$\therefore \overline{AC}^2 = \overline{AB}^2 + \overline{CB}^2.$$

A reproduction of the relations established in § 73.

Note. — The circumference on AC as a diameter will pass through B.

Exercises. — 1. Show how to construct a square equivalent in area to a given rectangle.

2. Show how to construct a rectangle of given side, the area of which shall equal a given square.

3. Construct a rectangle the area of which shall equal the difference of two given squares.

Remark. — When the square on a segment of a line equals the rectangle of two other segments, the side of the square is said to be a mean proportional between the sides of the rectangle.

AREAS OF SIMILAR TRIANGLES. 95

78. Problem. *To find the relation existing between the areas of similar triangles.*

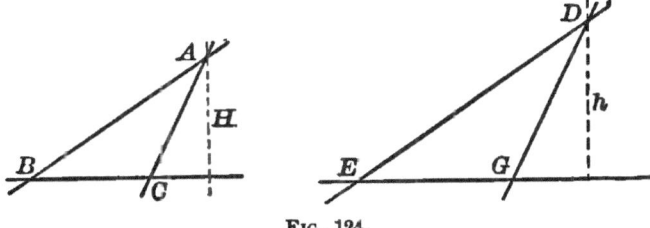

Fig. 124.

Let ABC and DEG represent the two triangles, H an altitude of ABC, and h the corresponding altitude of DEG.

$$\text{Area } ABC = \tfrac{1}{2}\,\overline{BC}\cdot H.$$
$$\text{Area } DEG = \tfrac{1}{2}\,\overline{EG}\cdot h.$$

Dividing member by member, we have:

$$\frac{\text{Area } ABC}{\text{Area } DEG} = \frac{\overline{BC}\cdot H}{\overline{EG}\cdot h}.$$

But
$$\frac{\overline{BC}}{\overline{EG}} = \frac{H}{h}. \quad \text{(Gen. Ex. 4, Chap. V.)}$$

$$\therefore \frac{\text{Area } ABC}{\text{Area } DEG} = \frac{H^2}{h^2}.$$

From Chapter IV. we know that the altitudes of similar triangles are proportional to any other corresponding lines that exist or may be drawn.

If B and b represent corresponding sides,

$$\frac{H}{h} = \frac{B}{b}.$$

By squaring both members,

$$\frac{H^2}{h^2} = \frac{B^2}{b^2}.$$

$$\therefore \frac{\text{Area } ABC}{\text{Area } DEG} = \frac{B^2}{b^2}.$$

And in the same way may be shown to have the same ratio as the squares on *any* corresponding lines. Q. E. F.

79. Problem. *To find the relation existing between the areas of similar polygons.*

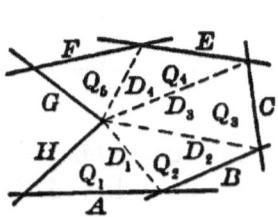 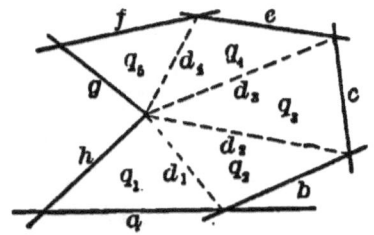

Fig. 125.

Let A and a represent two corresponding sides, B and b other corresponding sides, and so on about the two figures. The polygons being similar, the triangles formed by drawing corresponding diagonals will be similar (Gen. Ex. 2, Chap. V.). Represent the diagonals and areas as indicated in the figures.

$$\frac{Q_1}{q_1} = \frac{D_1^2}{d_1^2}; \quad \frac{Q_2}{q_2} = \frac{D_2^2}{d_2^2}; \quad \frac{Q_3}{q_3} = \frac{D_3^2}{d_3^2}; \quad \frac{Q_4}{q_4} = \frac{D_4^2}{d_4^2}; \quad \frac{Q_5}{q_5} = \frac{G^2}{g^2}.$$

The second members of these equations are all equal to each other.

$$\therefore \frac{Q_1}{q_1} = \frac{D_1^2}{d_1^2}; \quad \frac{Q_2}{q_2} = \frac{D_1^2}{d_1^2}; \quad \frac{Q_3}{q_3} = \frac{D_1^2}{d_1^2}; \quad \frac{Q_4}{q_4} = \frac{D_1^2}{d_1^2}; \quad \frac{Q_5}{q_5} = \frac{D_1^2}{d_1^2}.$$

These equations may be put in the form:

$$Q_1 d_1^2 = q_1 D_1^2.$$
$$Q_2 d_1^2 = q_2 D_1^2.$$
$$Q_3 d_1^2 = q_3 D_1^2.$$
$$Q_4 d_1^2 = q_4 D_1^2.$$
$$Q_5 d_1^2 = q_5 D_1^2.$$

AREAS OF SIMILAR POLYGONS. 97

Or, adding member to member:
$$(Q_1 + Q_2 + Q_3 + Q_4 + Q_5)d_1^2 = (q_1 + q_2 + q_3 + q_4 + q_5)D_1^2$$
Or,
$$\frac{Q_1 + Q_2 + Q_3 + Q_4 + Q_5}{q_1 + q_2 + q_3 + q_4 + q_5} = \frac{D_1^2}{d_1^2}.$$

If P and p represent the areas of the respective polygons,
$$\frac{P}{p} = \frac{D_1^2}{d_1^2}.$$

But the ratio $\frac{D_1^2}{d_1^2}$ is the same as the ratio of the squares of any other corresponding lines.

The method of treatment here employed is in no way dependent upon the number of angles of the polygons and so may be extended to polygons having any number of angles.

Thus we are not in danger of drawing general conclusions from special cases.

Hence the theorem:

The areas of similar polygons are proportional to the squares on any corresponding lines. Q. E. F.

80. PROBLEM. *To find the relation that exists between the areas of two triangles which have an angle in each equal.*

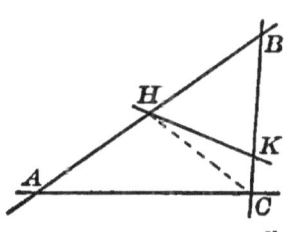

FIG. 126.

Let ABC and DEF represent the two triangles, having the interior angles at B and E equal.

98 ELEMENTS OF GEOMETRY.

Place the latter so that they will coincide.

Let BHK be the position taken by DEF.

Draw the auxiliary line HC.

$$\frac{\triangle HBK}{\triangle HBC} = \frac{\overline{BK}}{\overline{BC}}, \qquad \text{(Ex. 4, § 60.)}$$

$$\frac{\triangle HBC}{\triangle ABC} = \frac{\overline{BH}}{\overline{BA}},$$

$$\frac{\triangle HBK \cdot \triangle HBC}{\triangle HBC \cdot \triangle ABC} = \frac{\overline{BK} \cdot \overline{BH}}{\overline{BC} \cdot \overline{BA}}.$$

Dividing numerator and denominator of the first member by $\triangle HBC$, we have

$$\frac{\triangle HBK}{\triangle ABC} = \frac{\overline{BK} \cdot \overline{BH}}{\overline{BC} \cdot \overline{BA}}; \text{ or,}$$

The areas of two triangles having an angle in each equal, are proportional to the rectangles formed by the including sides. Q. E. F.

Exercises. — 1. Draw the auxiliary line from A to K and determine the relation.

2. ABC is any triangle, DE is parallel to AC, DC is a diagonal of the trapezoid.

Show that $\triangle BDC$ is a mean proportional between $\triangle BDE$ and $\triangle BAC$.

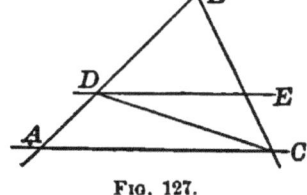

Fig. 127.

3. Through the vertex of a triangle to draw a straight line that shall separate the triangle into parts that shall have any given ratio $\frac{m}{n}$.

AREAS OF SIMILAR POLYGONS.

81. Problem. *Having given the three sides of a triangle, determine the segments into which the base is separated by a perpendicular from the opposite vertex.*

Representing the various segments by single letters,* we have:

$$c^2 = a^2 + b^2 - 2\,by, \quad (\S\ 76)$$
$$2\,by = a^2 + b^2 - c^2,$$
$$y = \frac{a^2 + b^2 - c^2}{2\,b}, \qquad (1)$$
$$x = b - y. \qquad (2)$$

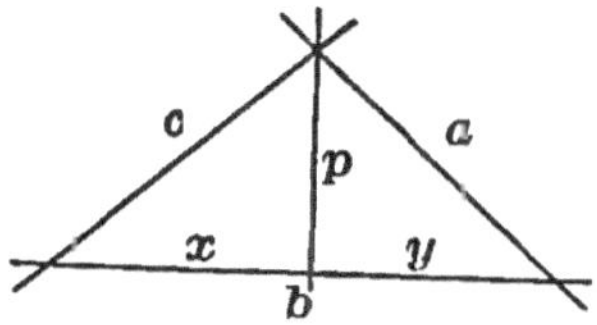

Fig. 128.

Equations (1) and (2) determine the segments.
If the perpendicular fall without the triangle, we have:

$$c^2 = a^2 + b^2 + 2\,by, \qquad (\S\ 75)$$
$$c^2 - a^2 - b^2 = 2\,by,$$
$$y = \frac{c^2 - a^2 - b^2}{2\,b}. \qquad (1)$$
$$x = b + y.$$

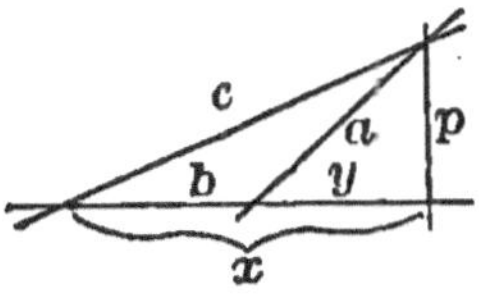

Fig. 129.

The student will observe the oneness of these apparently different results. Q. E. F.

82. Problem. *Find an expression for the median of a triangle in terms of the sides.*

$$a^2 = m^2 + \left(\frac{b}{2}\right)^2 + 2\left(\frac{b}{2}\right)q; \qquad (\S\ 75)$$

$$c^2 = m^2 + \left(\frac{b}{2}\right)^2 - 2\left(\frac{b}{2}\right)q; \qquad (\S\ 76)$$

---

*Note.—Portions of geometric figures are frequently represented by a single letter, when misunderstanding regarding the intent is not likely to arise from such representation.

$$a^2 + c^2 = 2m^2 + 2\left(\frac{b}{2}\right)^2 \qquad (t)$$

$$= 2m^2 + \frac{b^2}{2},$$

$$2m^2 = a^2 + c^2 - \frac{b^2}{2},$$

$$2m^2 = \frac{2a^2 + 2c^2 - b^2}{2},$$

$$m^2 = \frac{2a^2 + 2c^2 - b^2}{4}.$$

$$m = \tfrac{1}{2}\sqrt{2a^2 + 2c^2 - b^2}. \qquad (u)$$

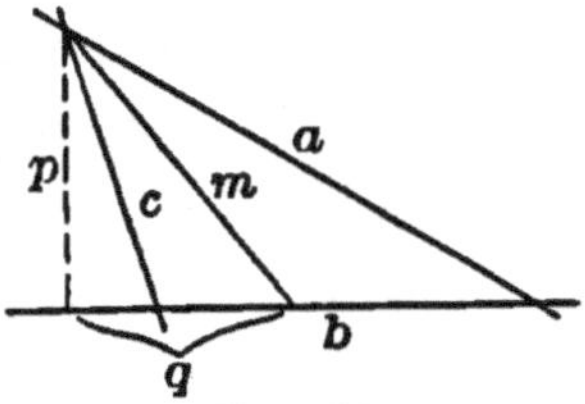

FIG. 180.

Equation (t) expresses the relation in one form. (u) is simply a solution of equation (t) with respect to m. Q. E. F.

83. PROBLEM. *Find an expression for the altitude of a triangle in terms of the sides.*

Let ABC represent a triangle; the perpendicular, p, separating the base into the segments x and y.

$$p^2 = a^2 - y^2,$$

$$p^2 = (a+y)(a-y).$$

But by § 81, $y = \dfrac{a^2 + b^2 - c^2}{2b}.$

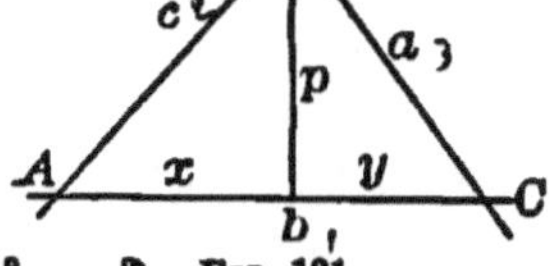

FIG. 181.

$$\therefore p^2 = \left(a + \frac{a^2 + b^2 - c^2}{2b}\right)\left(a - \frac{a^2 + b^2 - c^2}{2b}\right),$$

$$p^2 = \left(\frac{2ab + a^2 + b^2 - c^2}{2b}\right)\left(\frac{2ab - a^2 - b^2 + c^2}{2b}\right),$$

$$p^2 = \left(\frac{(a+b)^2 - c^2}{2b}\right)\cdot\left(\frac{-(a-b)^2 + c^2}{2b}\right),$$

$$p^2 = \frac{(a+b+c)(a+b-c)(a+c-b)(b+c-a)}{4b^2},$$

METRICAL RELATIONS. 101

A more condensed expression may be produced by substituting the perimeter s for $a + b + c$. If $a + b + c = s$,

$$a + b - c = s - 2c,$$
$$a + c - b = s - 2b,$$
$$b + c - a = s - 2a.$$

$$\therefore p^2 = \frac{s(s-2c)(s-2b)(s-2a)}{4b^2},$$

or, $$p = \frac{1}{2b}\sqrt{s(s-2c)(s-2b)(s-2a)}.\quad \text{Q. E. F.}$$

Exercise. — Show that the formula will be the same if the perpendicular fall without the triangle.

84. Problem. *Find an expression for the area of a triangle in terms of its sides.*

$$\text{Area} = \tfrac{1}{2} bp,$$

$$\text{Area} = \tfrac{1}{2} b \cdot \frac{1}{2b}\sqrt{s(s-2a)(s-2b)(s-2c)}, \quad (\S\ 83)$$

$$\text{Area} = \tfrac{1}{4}\sqrt{s(s-2a)(s-2b)(s-2c)}.$$

If the $\tfrac{1}{4}$ be introduced under the radical sign,

$$\text{Area} = \sqrt{\frac{s(s-2a)(s-2b)(s-2c)}{16}},$$

$$\text{Area} = \sqrt{\frac{s}{2}\left(\frac{s}{2}-a\right)\left(\frac{s}{2}-b\right)\left(\frac{s}{2}-c\right)}.$$

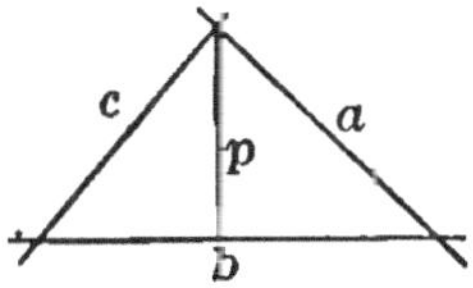

Fig. 182.

Exercise. — Find the area of a triangular piece of land, the three sides of which are: 32.93 chains, 48.26 chains, and 51.48 chains.

GENERAL EXERCISES.

1. Find expressions in terms of the sides, for each of the segments into which an angle-bisector separates the opposite side of a triangle.

2. The sides of a triangle are 8, 12, and 15. Has the triangle an obtuse angle?

3. Show that the areas of two triangles having an angle in one the supplement of an angle in the other are to each other as the rectangles on the including sides.

4. Show that the sum of the squares of the sides of any quadrangle equals the sum of the squares of the diagonals *plus* four times the square of the line joining the middle points of the diagonals.

5. Show that in a trapezoid, the sum of the squares of the diagonals equals the sum of the squares on the non-parallel sides *plus* twice the rectangle on the parallel sides.

6. Construct a square equal to the sum of two given squares.

7. Show that *four* times the sum of the squares on the medians of a triangle equals *three* times the sum of the squares on the sides.

8. If any point P in the plane of a triangle be joined with the three vertices A, B, C and the point of intersection I of the medians, $\overline{PA}^2 + \overline{PB}^2 + \overline{PC}^2 = \overline{AI}^2 + \overline{BI}^2 + \overline{CI}^2 + 3\,\overline{PI}^2$.

9. Show how to draw a line parallel to any side of a triangle so as to bisect the area.

10. Show how to draw a line through any point in one side of a triangle so as to bisect the area.

11. Show how to draw a line parallel to a given line so as to bisect the area of a given triangle.

CHAPTER VII.

85. Theorem. *If two chords intersect within a circle, the rectangle on the segments of one chord equals the rectangle on the segments of the other.*

Let AB and CD represent two chords intersecting at I. Draw the auxiliary lines AC and DB. The vertical angles at I are equal;

$$\angle CAI = \angle BDI,$$

each being subtended by the arc CKB.

$$\angle ACI = \angle DBI,$$

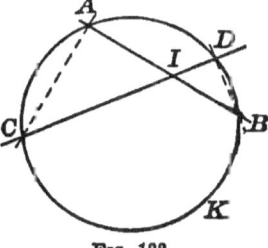

Fig. 133.

each being subtended by the arc AD.

∴ $\triangle AIC$ is similar to $\triangle DIB$.

Then $\dfrac{AI}{ID} = \dfrac{CI}{IB}$ or $\overline{AI} \times \overline{IB} = \overline{CI} \times \overline{ID}$.   Q. E. D.

Exercise. — Show how to construct a square that shall be equivalent to a given rectangle, the sides of which are a and b. If two chords be so drawn that the segments of one were adjacent sides of a rectangle and the segments of the other were equal, by the theorem we would have a rectangle equivalent to a square.

We know (§ 45) that if a diameter be drawn perpendicular to a chord it will bisect the chord.

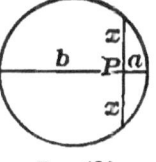

Fig. 134.

Hence if a circumference be constructed with $(a + b)$ as a diameter, and at the common extremity of a and b

104 ELEMENTS OF GEOMETRY.

a perpendicular chord be constructed, half the chord will be the side of the required square, since $ab = x^2$.

The equation $x^2 = ab$ is thus solved geometrically and exactly.

86. THEOREM. *If two secants intersect without the circumference, the rectangle on the distances from the common point to the two intersections with the circumference in one case will be equal to the rectangle similarly formed in the other.*

The two $\triangle\ DAC$ and BEC are similar. $\angle\ C$ is common. $\angle\ CDA = \angle\ CBE$ (being inscribed in the same segment). Therefore the third angles are equal.

Then, by § 64,

$$\frac{CE}{CA} = \frac{CB}{CD}, \text{ or } \overline{CE} \times \overline{CD} = \overline{CA} \times \overline{CB}. \qquad \text{Q. E. D.}$$

FIG. 135.

Exercises. — 1. Having given a rectangle, construct an equivalent rectangle that shall have a given side.

2. If one secant remain stationary, and the other rotate about the common point C until it become a tangent, we shall have a secant and a tangent.

Show that $\overline{CT}^2 = \overline{CA} \times \overline{CB}$.

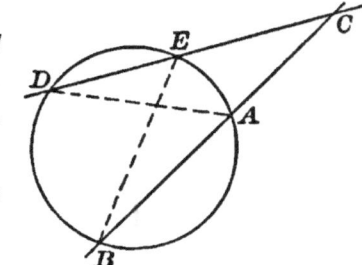

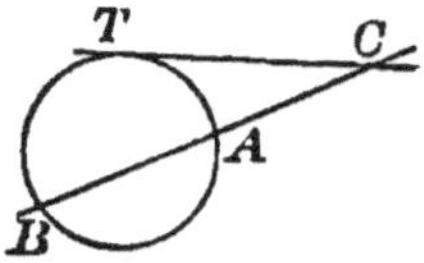

FIG. 136.

3. Use Exercise 2 to construct a square equivalent to a given rectangle.

4. Construct a square equivalent to a given triangle.

5. $x = \dfrac{abc}{de}$. Construct x.

Suggestion. — Put $x = \dfrac{a}{d} \cdot \dfrac{bc}{e}$; construct $q = \dfrac{bc}{e}$, then construct $x = \dfrac{aq}{d}$.

CHORDS AND TANGENTS. 105

87. Problem. *Find an expression for the bisector of an angle of a triangle in terms of the including sides and the segments of the third side.*

Circumscribe a circle about the triangle, and draw the auxiliary line AK. $\triangle KAD$ and CHD are similar.

$$\frac{b+g}{c} = \frac{a}{b},$$

$$b^2 + bg = ac,$$

$$b^2 = ac - bg.$$

But by § 85, $bg = mn.$

$\therefore b^2 = ac - mn.$

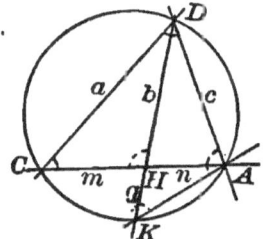

Fig. 137.

If it be desired to find m and n in terms of the sides a, c, and d of the triangle,

$$m + n = d \text{ and } \frac{m}{n} = \frac{a}{c}$$

The solution of this pair of equations will give m and n, in terms of a, c, and d. Q. E. F.

88. Problem. *To find a relation between the sides and the diagonals of a quadrangle inscribed in a circle.*

Let a, b, e, and d represent the sides of the inscribed quadrangle, and m, n, p, and q the segments of the diagonals.

The $\triangle mbq$ and dpn are similar (being mutually equiangular).

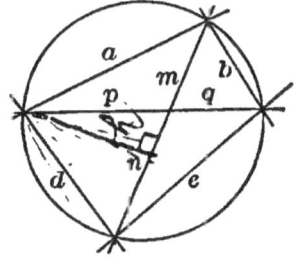

Fig. 138.

Then, $\quad \dfrac{b}{m} = \dfrac{d}{p}; \; \dfrac{bd}{m} = \dfrac{d^2}{p}; \; bd = \dfrac{md^2}{p}.$ (1)

Because the △ enq and apm are similar (being mutually equiangular),

$$\frac{e}{n} = \frac{a}{p}; \quad \frac{ae}{n} = \frac{a^2}{p}; \quad ae = \frac{na^2}{p}. \tag{2}$$

Adding the members of equations (1) and (2), we have

$$bd + ae = \frac{md^2 + na^2}{p}. \tag{3}$$

Equation (3) expresses a relation between the sides and segments of the diagonals, but a more convenient relation may be obtained by transforming the second member.

By §§ 75 and 76, representing the perpendicular projection of p on the other diagonal by h,

$$d^2 = p^2 + n^2 \pm 2nh, \tag{4}$$
$$a^2 = p^2 + m^2 \mp 2mh. \tag{5}$$

Multiplying both members of (4) by m, and both members of (5) by n, we have

$$md^2 = mp^2 + mn^2 \pm 2mnh,$$
$$na^2 = np^2 + m^2n \mp 2mnh,$$
$$md^2 + na^2 = (m+n)p^2 + (m+n)mn.$$

But $\quad mn = pq.$ (§ 85)

$$\therefore md^2 + na^2 = (m+n)p^2 + (m+n)pq$$
$$= (m+n)(p+q)p.$$

Substituting in (3), we have

$$bd + ae = \frac{(m+n)(p+q)p}{p} = (m+n)(p+q).$$

Which expressed in words is:

The sum of the rectangles on the opposite sides equals the rectangle on the diagonals. Q. E. F.

CHORDS AND TANGENTS. 107

89. Problem. *Show how to construct a square that shall have the same ratio to a given square as two given straight line segments, h and k.*

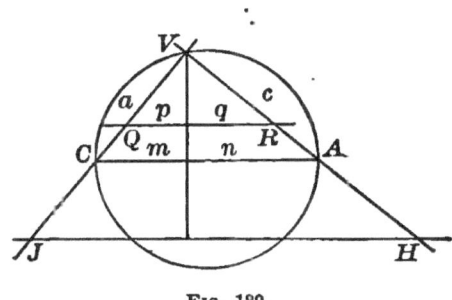

Fig. 139.

Analysis. — *If* the required square were known, and if sides of the given and required squares were placed at right angles to each other and so that they had a common extremity, by joining their other extremities a right triangle would be formed; and if from the vertex of the right angle a perpendicular were drawn to the hypothenuse, it would separate the latter into segments proportional to the squares on the corresponding sides. [§ 77 (c).]

Construction. — Remembering that if a circumference be constructed with the hypothenuse of a right triangle as its diameter, it will pass through the vertex, we have, taking m and n of convenient length and so that $\dfrac{m}{n} = \dfrac{h}{k}$, erecting a perpendicular at their common extremity and forming the $\triangle CVA$:

$$\frac{a^2}{c^2} = \frac{m}{n}.$$

If a side of the given square be less than a, lay off VQ

equal to a side of the given square and draw QR ∥ to CA. VR will be a side of the required square.

$$\frac{\overline{VR}^2}{\overline{VQ}^2} = \frac{q}{p} = \frac{n}{m} = \frac{k}{h}.$$

If a side of the given square be greater than a, lay off VJ equal to that side.

Draw JH ∥ to CA. VH will be a side of the required square. Q. E. F.

90. PROBLEM. *Show how to construct a common tangent to two given circumferences.*

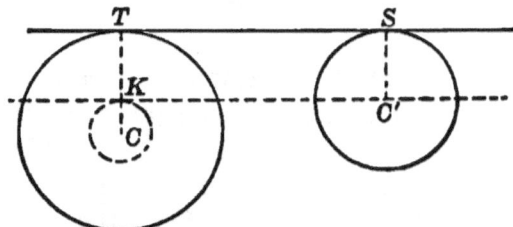

FIG. 140.

Let the circles whose centres are C and C' represent the two circles, so chosen that they do not form a special case.

Analysis. — If TS were the required line tangent to the two circumferences at T and S, and *if* we should draw $C'S$ and CT, they would both be perpendicular to the tangent, and hence parallel to each other.

If through C' a line were drawn parallel to ST, the figure $C'KTS$ would be a rectangle.

If with C as a centre, and CK as radius, a circle were constructed, $C'K$ would be tangent to the constructed circumference at the point K, because perpendicular to a radius at its extremity. The radius of this constructed

CHORDS AND TANGENTS. 109

circle would be the difference of the radii of the given circles.

Construction. — With a radius equal to the difference of the radii of the two circles, and with C as a centre, construct a circle.

Through C' draw $C'H$, a tangent to the newly constructed circle (Prob. 1, Gen'l Ex., Chap. IV.). Draw CH, intersecting the larger circumference at J. Through C' draw $C'M$ parallel to CJ. Each will be perpendicular to $C'H$, and a line joining the points M and J will be parallel to $C'H$, and hence perpendicular to $C'M$ and CJ. In each case being perpendicular to a radius at its extremity, it will be tangent to each circle, and so a common tangent.*

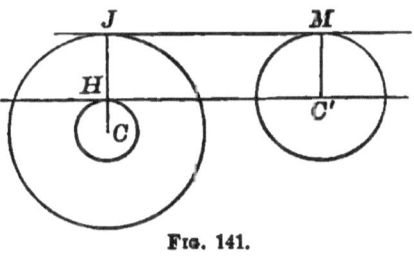

Fig. 141.

Discussion. — Since through any point outside a circumference two tangents may be drawn, we have a double construction as indicated in the accompanying figure.

But it appears that a tangent might be drawn that should pass between the two circumferences.

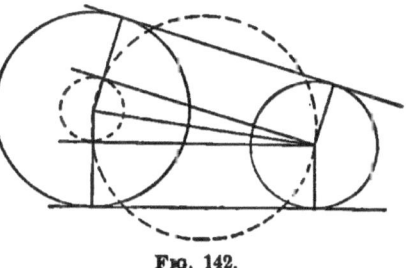

Fig. 142.

Analysis.—If BD were such a tangent, and if we should draw lines from C and

---

* NOTE. — The student should construct the figure as the reading progresses, not in advance of it.

110   ELEMENTS OF GEOMETRY.

C' to the points of tangency, they would be perpendicular to the same line, and so parallel. *If* through C', $C'E$ were drawn parallel to DB, it would be perpendicular to CE. *If* then with C as a centre and a radius CE, an auxiliary circle were drawn, $C'E$ would be tangent to it at the point E.

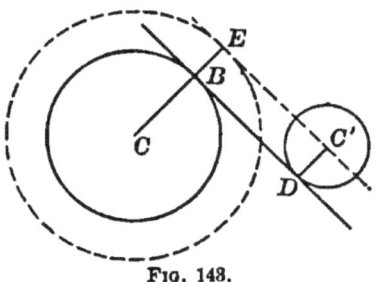

Fig. 143.

The radius of the auxiliary circle would be the sum of the radii of the given circles.

Construction. — With C as a centre and a radius equal to the sum of the radii of the given circles, construct a circle. Through C' draw a tangent $C'Q$ to the auxiliary circumference (Gen'l Ex. 1, Chap. IV.). Draw CQ, and through C' draw $C'R$ ∥ to CQ. Each will be perpendicular to $C'Q$. Draw

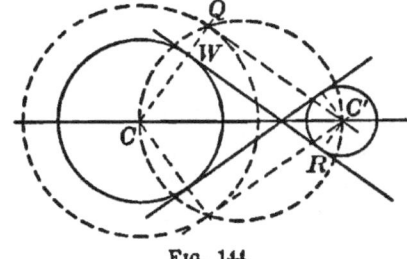

Fig. 144.

WR; it will be perpendicular to a radius at its extremity in each case, and so tangent to each circle.

Discussion. — Since from the point C' two tangents may be drawn to the auxiliary circle, a double construction may be had, and there will be two tangents which pass between the given circles.

Further Discussion. — It is now seen that there may be *four* tangents to two given circumferences, two called **external** tangents and two **internal**. If taken in the general position, the radii remaining constant, and one of the centres be moved toward the other one, the tangents will change the angles that they make with each other, and when the

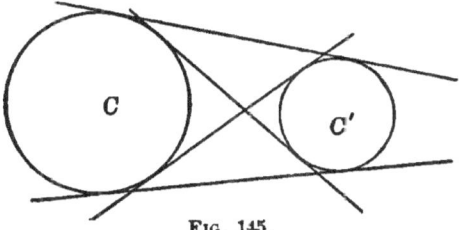

Fig. 145.

circumferences become externally tangent, the interior tangents will coincide and will form but one tangent.

In this case, which is a special one, two tangents have become coincident, and there will be *three* real tangents.

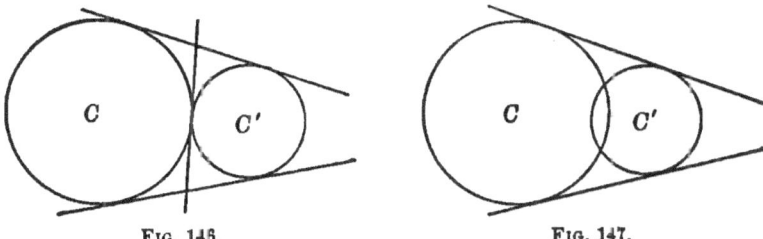

Fig. 146.   Fig. 147.

Further motion in the same direction causes the circumferences to intersect, the internal tangents are said to become *imaginary*, and there will be *two* real tangents.

After a sufficient motion to bring the circumferences so that one shall be tangent to the other internally, the exterior tangents will have come to coincide. We say that there are *two imaginary* tangents and *two real*, but coincident, tangents.

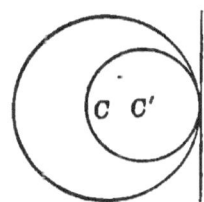

Fig. 148.

A further motion in the same direction will place one circumference entirely within the other, and all the tangents become imaginary. In other words, the geometric construction is impossible.

Note. — The analysis, construction, and discussion of the above problem have been set forth thus thoroughly, for the purpose of exhibiting to the student the *method* that is to be pursued in every such problem. The student should bear in mind the fact that the *construction* is the *inverse* of the *analysis*, and that each requires its own figure; and that the *discussion* determines the *limitations* of the problem.

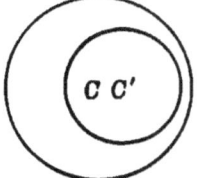

Fig. 149.

Henceforth problems of this character will not in the text be presented in such detail, but the student is expected to furnish the full detail.

91. Problems. — 1. Show that the exterior and the interior tangents to two circles intersect on the line of centres.

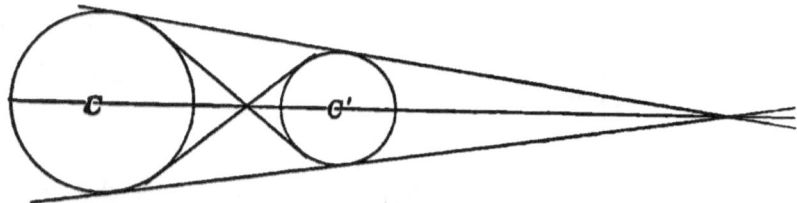

Fig. 150.

2. Show that in either case the distances from the centres to the point of intersection have the same ratio as the radii of the circles.*

3. If a third circle be tangent to two given circles, the line through the points of tangency will pass through the external centre of similitude, if both the given circles are externally or internally tangent to the third circle.

---

* Note. — The line of centres is said to be divided externally and internally in the same ratio. These points of division are called the external and internal centres of similitude.

CHORDS AND TANGENTS. 113

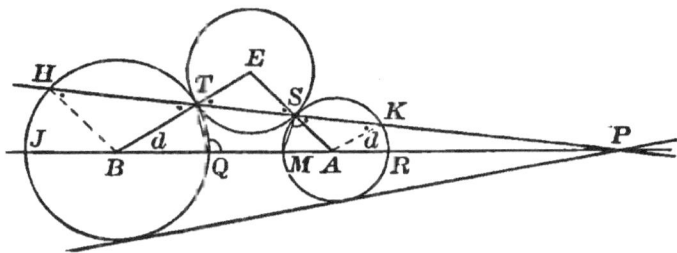

Fig. 151.

4. Show that in the last figure, $\overline{PS} \times \overline{PT} = \overline{PM} \times \overline{PQ}$.

5. Show how to construct a circumference tangent to a given circumference at a given point and also tangent to another given circumference.

6. Show that if the centres of the circles escribed to a triangle be joined, a triangle will be formed, the sides of which will pass through the vertices of the given triangle.

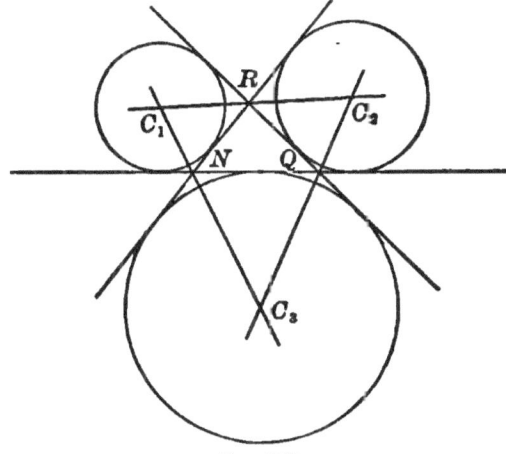

Fig. 152.

7. Show that a perpendicular from the centre of either escribed circle to the line of centres of the other two will intersect it at a vertex of the given triangle and will bisect the interior angle of the triangle at that vertex.

I

114   ELEMENTS OF GEOMETRY.

8. Show that the perpendicular mentioned in the foregoing exercise will pass through the centre of the *inscribed* circle.

9. Having given the centres of the escribed circles to a triangle, construct the triangle.

10. Apply the "Reductio ad Absurdum" (sometimes called *indirect*) method to the establishment of the fact that if three circles are tangent to each other, two and two, the common tangents will be concurrent.

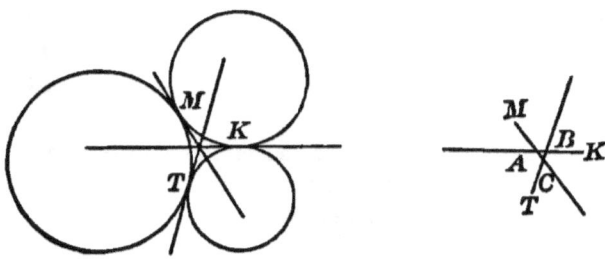

Fig. 153.

If the common tangents are *not* concurrent, they will intersect so as to form a triangle as indicated in the supplementary figure.

By Gen'l Ex. 6, Chap. IV.,

$$AM = AK$$
$$BK = BT$$
$$CT = CM$$
$$\overline{AM + BK + CT = AK + BT + CM}$$

But each term in the first member is less than a term in the second member.

$$AM < CM$$
$$BK < AK$$
$$CT < BT$$
$$\overline{AM + BK + CT > CM + AK + BT}$$

CHORDS AND TANGENTS.   115

The supposition that the tangents intersect so as to form a triangle is thus seen to be an erroneous one.

The tangents *may* all pass through one point, and *cannot* pass so as to form a triangle. Hence they do all pass through one point.   Q. E. D.

11. Show that if three circumferences intersect, the common chords will all pass through one point.

12. It has been said that the three common tangents to three tangent circumferences bisect the angles of the triangle formed by joining the points of tangency. Is it true?

FIG. 154.

13. Show how to separate a given segment of a line into two parts such that the ratio of the whole segment to the larger part shall equal the ratio of the larger part to the smaller part.

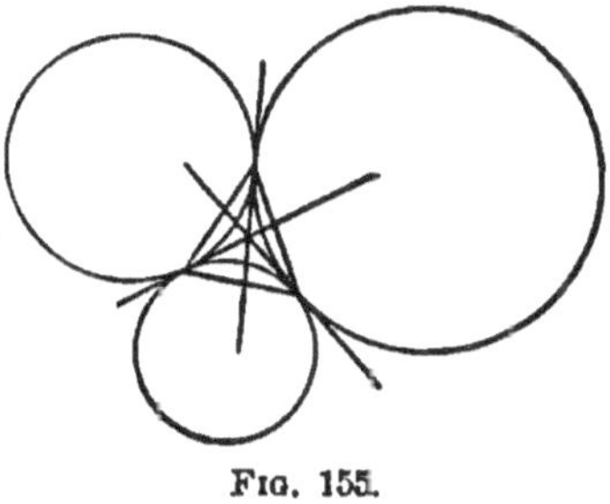

If we had the segment separated as desired, and if we let a represent the segment, x the larger part, and $a - x$ the smaller part, we should have:
$$\frac{a}{x} = \frac{x}{a-x},$$
or, $a^2 - ax = x^2$;
$x^2 + ax = a^2$, $x(a+x) = a^2$.

FIG. 155.

The form of the relation suggests that a secant and a tangent to a circumference might be used to express the relation geometrically.

If PQ represent the segment a, and if we have a circle tangent at Q, any secant from P will be so divided that the rectangle on the distances from P to the concave and convex arcs will equal a^2.

But the difference of the distances from P to the concave and convex arcs must be a; since one of the distances is to be x and the other *is to be* $x + a$.

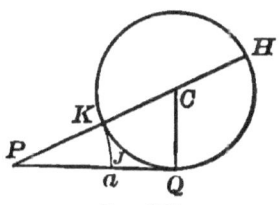

FIG. 156.

If the constructed circle have for its diameter a, and the secant be drawn through the centre, it will fulfil the required conditions.

Hence at one extremity of the given segment, as at Q, erect a perpendicular equal to $\frac{a}{2}$. With $\frac{a}{2}$ as a radius construct a circumference tangent to PQ at Q. Draw the secant PC. PK will represent x or the larger part into which a is to be separated.

If it be desired to lay that part off on PQ, with PK as a radius and P as a centre, construct an arc that will intersect PQ at J.

Note. — When a segment of a line is thus separated it is said to be divided into *mean and extreme ratio;* x is the mean, and the extremes are a and $a - x$. When written in extended form it is: $a : x :: x : a - x$.

CHAPTER VIII.

Definition. A regular polygon is one the angles of which are equal to each other and the sides are equal to each other.

92. Theorem. *A circumference may be circumscribed about any regular polygon.*

Let the accompanying figure represent a regular polygon. The bisectors of two adjacent angles will meet at some point, as C, forming the isosceles $\triangle CAB$.

If C be joined with D,

$$\triangle CBD = \triangle CBA.$$

$\therefore \triangle CBD$ is isosceles, $CB = CD$, and the angle of the polygon at D is bisected.

If C be joined with the other vertices, triangles will be formed each equal to CAB.

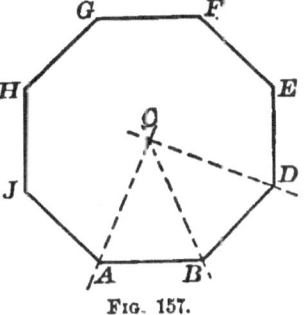

Fig. 157.

Therefore the distances from the point of intersection of the bisectors of a pair of adjacent angles from all the vertices, is the same, and if, with this point as a centre and a radius equal to the distance from this point to any vertex, a circumference be described, it will pass through each vertex. Q. E. D.

Remarks.— The centre of the circumscribed circle, which is also the centre of the inscribed circle, is called the **centre of the polygon**.

118   ELEMENTS OF GEOMETRY.

The radius of the circle circumscribed about a regular polygon is called the **radius** of the polygon.

The radius of the circle inscribed in a regular polygon is called the **apothem**.

Exercises. — 1. Show that perpendiculars from C to the sides of the polygon will all be equal, and that a circumference may be inscribed within a regular polygon.

2. Show that an inscribed equilateral polygon will be equiangular, and hence regular.

3. Show that an inscribed equiangular polygon may not be regular. Discuss.

4. Show that an equiangular polygon circumscribed about a circumference is regular.

5. Show that regular polygons of the same number of sides are similar figures.

93. Theorem. *If a regular polygon be inscribed in a circumference and then tangents be drawn parallel to the sides of the inscribed polygon, a circumscribed polygon will be formed that will be similar to the inscribed polygon.*

Let DA, AB, and BE represent three adjacent sides of the regular inscribed polygon. Let CF, CG, and CH be the perpendicular bisectors of these sides. They will also bisect the subtended arcs at O, M, and J.

If at these points, tangents be drawn, they will be parallel to the sides of the inscribed polygon.

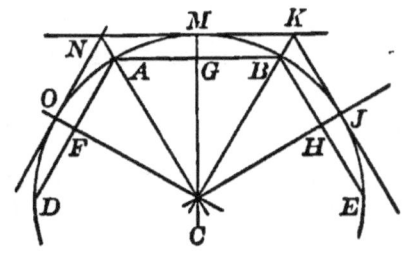

Fig. 158.

$$\angle ONM = \angle FAG,$$
$$\angle MKJ = \angle GBH,$$
etc.   etc.

The **angles** of the circumscribed polygon equal those

of the inscribed, since the sides are parallel and extend in the same direction.

Hence (Ex. 4, § 92) the circumscribed polygon is regular. But each side of the latter corresponds to a side of the inscribed polygon. Therefore they have the same number of sides, and being regular, are similar. Q. E. D.

94. THEOREM. *If at the vertices of a regular inscribed polygon, tangents be drawn, a circumscribed polygon will be formed which will be similar to the inscribed one.*

Let A, B, C, D, etc., represent the vertices of the inscribed polygon, at which tangents are drawn, intersecting at E, F, G, etc.

The $\triangle ABE$, BCF, and CDG are isosceles and are equal to each other (each having a side and two adjacent angles in each equal).

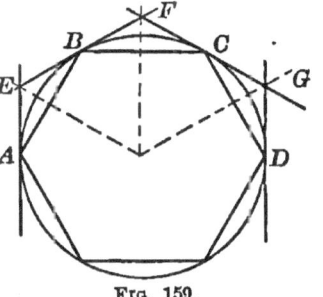

FIG. 159.

Therefore the angles of the circumscribed polygon are all equal to each other.

Each side of the circumscribed polygon, being the sum of two segments (each of which is equal to any one segment, as BF), are equal to each other.

Hence the polygon constructed fulfils the requirements for regularity, and having the same number of sides as the inscribed polygon will be similar to it. Q. E. D.

Exercises. — 1. Show how to cut a square piece of board so as to get a regular octagon.

2. Show that if chords be drawn from the vertices of a regular inscribed polygon to the middle points of the subtended arcs, a new regular inscribed polygon of double the number of sides will be formed.

FIG. 160.

120 ELEMENTS OF GEOMETRY.

3. Show that if the alternate vertices of a regular polygon of an even number of sides be joined, a regular polygon of half the number of sides will be formed.

4. Construct regular polygons of 3, 4, 6, 8, 12, and 16 sides.

5. Show that the area of a regular polygon equals half the rectangle on its perimeter and apothem.

6. Show that the area of any polygon circumscribed about a circle equals half the rectangle on its perimeter and the radius of the circle.

95. THEOREM. *If a regular polygon be circumscribed about a circle, lines be drawn from the centre to the vertices, and tangents be drawn at the points where these lines intersect the circumference, a regular circumscribed polygon of double the number of sides will be formed.*

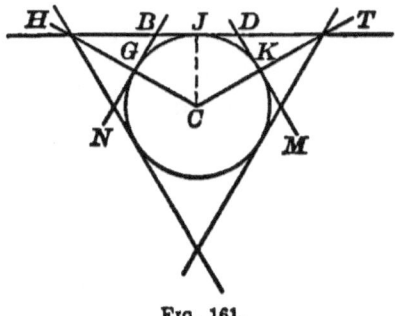

FIG. 161.

Let HT represent a side of a circumscribed regular polygon, G and K, points at which tangents are drawn as required.

$$MK = KD = DJ,$$
$$BJ = BG,$$
$$GB = DK$$

(because $\triangle HGB = \triangle DKT$).

$$\therefore MK = KD = DJ = JB = BG = GN.$$
$$\therefore MD = DB = BN.$$

The same may be shown for the other sides of the new figure. Therefore the new figure will be equilateral.

∴ △ *TKM*, *TKD*, *HGB*, and *HGN* are equal to each other, ∠ *TMK* = ∠ *TDK* = ∠ *HBG* = ∠ *HNG*.

The supplements of these several equal angles are interior angles of the polygon. Therefore the interior angles of the new polygon are all equal.

Hence the new polygon, which will have double the number of sides of the original polygon, will be both equilateral and equiangular: which are the conditions of regularity. Q. E. D.

VARIABLE AND LIMIT.

96. Definitions. A *variable* is a changing quantity; such as the time that has elapsed from some given epoch to the present, which is always changing.

That which represents the height of the ocean tide, the chord of a circle moving parallel to itself, or the speed of a bullet through the air, are further examples.

As a further illustration let perpendicular diameters be drawn in a circle. The point generating the circumference will be at changing distances from these diameters. If x represent the distance of P (any given point on the circumference) from one diameter, and y represent the distance from the other diameter, x and y will be variables. Incidentally *they are so related that* $x^2 + y^2 = r^2$.

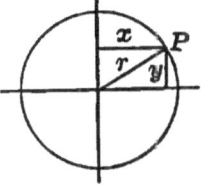

FIG. 162.

If there be a fixed quantity toward which a variable approaches, that fixed quantity is called a *limit*. The law governing a variable will determine whether it has a limit or not.

To illustrate: The repeating decimal .6666... will approach the fixed quantity $\tfrac{2}{3}$; the larger the number of terms considered, the nearer will the repeating decimal come to $\tfrac{2}{3}$. In this case the limit can be seen and named, although never reached by increasing the number of terms in the decimal.

If a secant AB rotate about the point A as a pivot, B will approach A. When B shall coincide with A, the line will have ceased to be a secant and will have become a tangent. If the point B, which is at a variable distance from A, shall continue to move along the circumference after having coincided with A, the line will again become a secant. The tangent is said to be the *limit* toward which the secant approaches as the second point of intersection approaches the first.

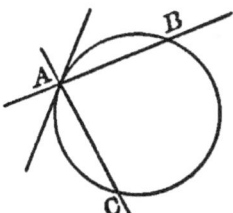

Fig. 163.

In this case the secant *reaches* its limit.

The series $1 + \tfrac{1}{2} + \tfrac{1}{4} + \tfrac{1}{8} + \tfrac{1}{16} + \cdots$ has for its limit the number 2, which we can see and can name, although it cannot be reached by any increase in the number of terms.

Let this problem be proposed: At what time between 2 and 3 o'clock will the hour and minute hands of the clock be together?

At 2 o'clock 10 minute spaces separate the hands. When the minute hand shall have reached the position primarily occupied by the hour hand, it will be behind the hour hand $\frac{1}{12}$ of the distance that first separated them. When the minute hand shall have reached the position occupied by the hour hand at the last accounting, the hour hand will be in advance $\frac{1}{12}$ of the distance that previously separated them.

Then the number of minute spaces to be passed over by the minute hand will be:

$$10 + \frac{10}{12} + \frac{10}{144} + \frac{10}{1728} + \frac{10}{(12)^4} + \frac{10}{(12)^5} + \cdots$$

That this series approaches a definite limit which can be named, we can determine by finding in another way the number of minute spaces passed over.

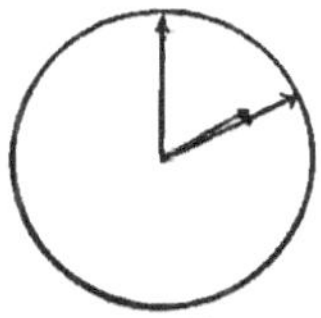

Let x represent the number of minutes after 2 that the hands are together (for they will be together). Then:

$$(x - 10)12 = x,$$
$$12x - 120 = x,$$
$$11x = 120, \quad x = 10\tfrac{10}{11}.$$

Fig. 164.

Therefore the series above written approaches $10\tfrac{10}{11}$ as its limit.

If θ represent one angle of a regular polygon of n sides,

$$\theta = 180° - \frac{360°}{n}.$$

As n increases, $\frac{360°}{n}$ will decrease, and θ approaches 180° as its limit.

If we have a series of numerical terms represented by $a + bh + ch^2 + dh^3 + eh^4 + \cdots$, the limit toward which this series will approach as we diminish the value of h will be a.

A limit may be of the same kind as the magnitude which approaches it, or it may be of a different kind.

Depending upon the nature of *variable* and *limit*, a *limit* may be reached, or it may not be reached.

There are two *axioms* relating to variables and limits which are of great importance.

Axiom I. *If the ratio of two variable quantities which approach limits always equals a fixed quantity, as the variables approach their limits, the* LIMITS *(or limiting values) will have the same ratio.*

Illustration. — $\dfrac{X_1}{Y_1} = \dfrac{a}{b}$; $\dfrac{X_2}{Y_2} = \dfrac{a}{b}$; $\dfrac{X_3}{Y_3} = \dfrac{a}{b}$, etc.

If X_1, Y_1, X_2, Y_2, etc., represent variables that at each stage have the same ratio, $\dfrac{a}{b}$, and these variables approach X and Y as their limits, $\dfrac{X}{Y} = \dfrac{a}{b}$.

Axiom II. *If two ratios, the terms of which are composed of variables approaching limits, are always equal to each other, as the limits are approached, they will be equal to each other at the limits.*

Illustration. — W_1, X_1, Y_1, Z_1, W_2, X_2, etc., represent limit-approaching variables, and at different halting-places of investigation we have

$$\dfrac{W_1}{X_1} = \dfrac{Y_1}{Z_1}, \quad \dfrac{W_2}{X_2} = \dfrac{Y_2}{Z_2}, \quad \dfrac{W_3}{X_3} = \dfrac{Y_3}{Z_3}, \quad \text{etc.}$$

Then if W, X, Y, and Z represent the limits approached,

$$\dfrac{W}{X} = \dfrac{Y}{Z}$$

VARIABLE AND LIMIT.

Exercises. — *Using the Limits Axioms,* —

1. Show that if a line be drawn parallel to a side of a triangle, it divides the other sides proportionally. (Suggestion: Approach the incommensurable limits through successive commensurable ratios.)

2. Show that in the same or equal circles, angles at the centre are proportional to the intercepted arcs.

3. Show that the areas of two rectangles are to each other as the products of base and altitude.

97. Theorem. *If a regular polygon be circumscribed about a circle, and a regular polygon of double the number of sides be constructed (as indicated in § 95), the latter will be of less area than the former, and of less perimeter.*

Using the figure of § 95, we see that if n represent the number of sides, when the number of sides is made $2n$, that n isosceles triangles (like DTM) will have been eliminated, and nothing having been added, the area will have been diminished.

Furthermore, with each elimination of an isosceles triangle, the base takes the place of the other two sides as a part of the perimeter, and therefore the perimeter is diminished.

Exercises. — 1. Show that the circumferences of two circles have the same ratio as their diameters.

2. Show that the areas of two circles have the same ratio as the squares of their radii.

Note. — The student will observe that at *each* doubling of the number of sides of the circumscribed polygon, the area will be diminished. The area of the circle is the limit toward which the area of the circumscribed polygon approaches as the number of sides is increased by doubling.

98. Theorem. *If a regular polygon be inscribed in a circle, and a regular polygon of double the number of sides be constructed (as indicated in Ex. 2, § 94), the latter will be of larger area than the former and of greater perimeter.*

Let AB represent a side of a regular inscribed polygon, and AD and DB represent sides of an inscribed polygon of double the number of sides.

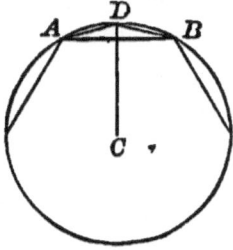

Fig. 165.

If n represent the number of sides in the given polygon, when the number of sides is doubled there will be added to the area of the polygon n isosceles triangles; and for each side of the polygon of n sides there will be substituted $AD + DB$ for AB, etc. Hence the theorem. Q. E. D.

Note. — The student will observe that at each doubling of the number of sides of the inscribed polygon the area will be increased no matter how many times the number of sides be doubled.

But the area of the polygon will always be less than the area of the circle, *toward which* we approach as the process goes on.

The circle is the *limit* toward which the inscribed polygon approaches as we increase the number of sides.

99. Problem. *To determine the law by which the areas of inscribed and circumscribed regular polygons approach the area of the circle as the sides are doubled in number.*

Let DB and MN represent sides of an inscribed and a circumscribed polygon of n sides.

DK will represent a side of the inscribed polygon of $2n$ sides, and HJ will represent a side of the circumscribed polygon of $2n$ sides.

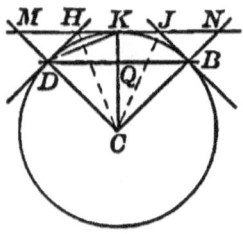

Fig. 166.

If a represent the area of the inscribed polygon, and A the area of the circumscribed polygon of n sides,

$$\triangle DCB = \frac{a}{n}, \quad \triangle DCQ = \frac{\triangle DCB}{2} = \frac{a}{2n},$$

$$\triangle MCN = \frac{A}{n}, \quad \triangle MCK = \frac{\triangle MCN}{2} = \frac{A}{2n}.$$

$$\therefore \frac{a}{A} = \frac{\triangle DCQ}{\triangle MCK}$$

$2n(\triangle DCK)$ equals the area of the inscribed polygon of $2n$ sides, and $2n(\triangle HJC)$, the area of the circumscribed polygon of $2n$ sides. If a' and A' represent these areas respectively,

$$\frac{a'}{A'} = \frac{2n(\triangle DCK)}{2n(\triangle HCJ)} = \frac{\triangle DCK}{\triangle HCJ}.$$

By § 80, the $\triangle DCK$ is a mean proportional to the $\triangle DCQ$ and MCK. Each multiplied by $2n$ would produce the entire areas of the polygons of which they are parts.

$$\therefore a' = \sqrt{aA}, \text{ or } \frac{1}{a'} = \frac{1}{\sqrt{aA}}; \qquad (1)$$

$$\frac{\triangle HCK}{\triangle MCK} = \frac{KH}{KM}.$$

But $\qquad 4n \cdot \triangle HCK = A'$
and $\qquad 2n \cdot \triangle MCK = A,$
or $\qquad 4n \cdot \triangle MCK = 2A.$

$$\therefore \frac{A'}{2A} = \frac{\triangle HCK}{\triangle MCK} = \frac{KH}{KM},$$

$$A' = \frac{\overline{KH}(2A)}{\overline{KM}} = \frac{2\overline{KH} \cdot A}{\overline{KH} + \overline{HM}}.$$

Taking the reciprocals of the last result, we have

$$\frac{1}{A'} = \frac{\overline{KH} + \overline{HM}}{2\,\overline{KH}\cdot A} = \frac{1}{2}\left(\frac{\overline{KH}+\overline{HM}}{\overline{KH}\cdot A}\right),$$

or $\quad \dfrac{1}{A'} = \dfrac{1}{2}\left(\dfrac{\overline{KH}}{\overline{KH}\cdot A} + \dfrac{\overline{HM}}{\overline{KH}\cdot A}\right) = \dfrac{1}{2}\left(\dfrac{1}{A} + \dfrac{\overline{HM}}{\overline{KH}\cdot A}\right).$

$$\left\{\begin{array}{l} \text{But}\quad \dfrac{a}{A} = \dfrac{\overline{CD}^2}{\overline{CM}^2} = \dfrac{\overline{CK}^2}{\overline{CM}^2} = \dfrac{\overline{KH}^2}{\overline{HM}^2},\ \S\ 69,\ \text{Prob. 6.} \\[2mm] \text{or}\quad a = \dfrac{\overline{KH}^2}{\overline{HM}^2}\cdot A\ \text{ or }\ \sqrt{a} = \dfrac{\overline{KH}}{\overline{HM}}\sqrt{A}. \end{array}\right\}$$

$$\frac{\overline{HM}}{\overline{KH}\cdot A} = \frac{\overline{HM}}{\overline{KH}\sqrt{A}\sqrt{A}} = \frac{1}{\left(\dfrac{\overline{KH}}{\overline{HM}}\sqrt{A}\right)\cdot\sqrt{A}} = \frac{1}{\sqrt{a}\sqrt{A}}$$

$$= \frac{1}{\sqrt{aA}}.\quad \therefore\ \frac{1}{A'} = \frac{1}{2}\left(\frac{1}{A}+\frac{1}{\sqrt{aA}}\right)\ (2).$$

The reciprocals of the areas are:

$$\frac{1}{a},\ \frac{1}{A},\ \frac{1}{\sqrt{aA}},\ \frac{1}{2}\left(\frac{1}{A}+\frac{1}{\sqrt{aA}}\right).$$

If then we know the reciprocals of the areas of inscribed and circumscribed polygons of n sides, the reciprocals of those of $2n$ sides may be determined by the above established relations.

100. *A Numerical Computation.* — If in a circle whose radius is unity we inscribe a square, and then circumscribe a square about it, we would have for the area of the inscribed square, $a = 2$; and for the area of the circumscribed square, $A = 4$; for the area of the inscribed octagon, $a' = \sqrt{8}$.

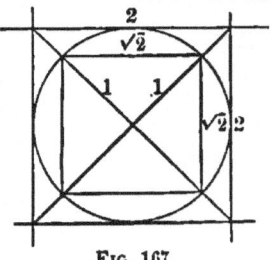

Fig. 167.

AREAS OF REGULAR POLYGONS.

$$\frac{1}{a} = \frac{1}{2} = .5. \quad \frac{1}{A} = \frac{1}{4} = .25.$$

$$\frac{1}{a_1} = \sqrt{.5 \times .25} = .3535533907 \ldots$$

$$\frac{1}{A_1} = \tfrac{1}{2}(.25 + .3535533907) = .3017766953 \ldots$$

$$\frac{1}{a_2} = \sqrt{\frac{1}{a_1} \cdot \frac{1}{A_1}} = .3266407412 \ldots$$

$$\frac{1}{A_2} = \frac{1}{2}\left(\frac{1}{A_1} + \frac{1}{a_2}\right) = .3142087182 \ldots$$

$$\frac{1}{a_3} = \sqrt{\frac{1}{a_2} \cdot \frac{1}{A_2}} = .3203644309 \ldots$$

$$\frac{1}{A_3} = \frac{1}{2}\left(\frac{1}{A_2} + \frac{1}{a_3}\right) = .3172865745 \ldots$$

$$\frac{1}{a_4} = \sqrt{\frac{1}{a_3} \cdot \frac{1}{A_3}} = .3188217885 \ldots$$

$$\frac{1}{A_4} = \frac{1}{2}\left(\frac{1}{A_3} + \frac{1}{a_4}\right) = .3180541815 \ldots$$

$$\frac{1}{a_5} = \sqrt{\frac{1}{a_4} \cdot \frac{1}{A_4}} = .3184377537 \ldots$$

$$\frac{1}{A_5} = \frac{1}{2}\left(\frac{1}{A_4} + \frac{1}{a_5}\right) = .3182459676 \ldots$$

$$\frac{1}{a_6} = \sqrt{\frac{1}{a_4} \cdot \frac{1}{A_5}} = .3183418462 \ldots$$

$$\frac{1}{A_6} = \frac{1}{2}\left(\frac{1}{A_5} + \frac{1}{a_6}\right) = .3182939069 \ldots$$

$$\frac{1}{a_7} = \sqrt{\frac{1}{a_6} \cdot \frac{1}{A_6}} = .3183178756 \ldots$$

K

$$\frac{1}{A_7} = \frac{1}{2}\left(\frac{1}{A_6} + \frac{1}{a_7}\right) = .3183058912\ldots$$

$$\frac{1}{a_8} = \sqrt{\frac{1}{a_7} \cdot \frac{1}{A_7}} = .3183118833\ldots$$

$$\frac{1}{A_8} = \frac{1}{2}\left(\frac{1}{A_7} + \frac{1}{a_8}\right) = .3183088872\ldots$$

$$\frac{1}{a_9} = \sqrt{\frac{1}{a_8} \cdot \frac{1}{A_8}} = .3183103853\ldots$$

$$\frac{1}{A_9} = \frac{1}{2}\left(\frac{1}{A_8} + \frac{1}{a_9}\right) = .3183096362\ldots$$

$$\frac{1}{a_{10}} = \sqrt{\frac{1}{a_9} \cdot \frac{1}{A_9}} = .3183100107\ldots$$

$$\frac{1}{A_{10}} = \frac{1}{2}\left(\frac{1}{A_9} + \frac{1}{a_{10}}\right) = .3183098234\ldots$$

$$\frac{1}{a_{11}} = \sqrt{\frac{1}{a_{10}} \cdot \frac{1}{A_{10}}} = .3183099170\ldots$$

$$\frac{1}{A_{11}} = \frac{1}{2}\left(\frac{1}{A_{10}} + \frac{1}{a_{11}}\right) = .3183098702\ldots$$

$$\frac{1}{a_{12}} = \sqrt{\frac{1}{a_{11}} \cdot \frac{1}{A_{11}}} = .3183098936\ldots$$

$$\frac{1}{A_{12}} = \frac{1}{2}\left(\frac{1}{A_{11}} + \frac{1}{a_{12}}\right) = .3183098819\ldots$$

$$\frac{1}{a_{13}} = \sqrt{\frac{1}{a_{12}} \cdot \frac{1}{A_{12}}} = .3183098877\ldots$$

$$\frac{1}{A_{13}} = \frac{1}{2}\left(\frac{1}{A_{12}} + \frac{1}{a_{13}}\right) = .3183098848\ldots$$

AREA OF A CIRCLE.

a_{13} and A_{13} are respectively the areas of regular polygons of 32768 sides, inscribed within and circumscribed about a circle; the square on the radius being the unit.

$$a_{13} = \frac{1}{.3183098877\ldots} = 3.1415926386\ldots$$

$$A_{13} = \frac{1}{.3183098848\ldots} = 3.1415926672\ldots$$

The area of the circle lies between the areas of a_n and A_n, and is nearer to either of them than they are to each other.

The same may be said of a_1 and A_1, a_2 and A_2, a_3 and A_3, etc., as far as the computations may be continued.

In our investigation we have carried the approximation to a_{13} and A_{13}, and find that they are the same for *seven* decimal places. The area of the circle will be the same for *seven* decimal places.

Letting K represent the area of a circle and R its radius, we will have:

$$K = (3.1415926\ldots)(R^2),$$

or
$$\frac{K}{R^2} = 3.1415926\ldots$$

The area of a circle is an exact thing and the square on the radius is an exact thing. The ratio of the two, by common consent, is represented by the Greek letter (π) and is an incommensurable number to which we may approach as near as time and patience will allow, but which we can never express in integers or their fractional parts.

NOTE.— The quadrature of a circle (*i.e.*, the determination of its area in terms of any given square) has demanded the attention of students of mathematics for 4000 years or more, and has had expended upon it a vast amount of time.

132 ELEMENTS OF GEOMETRY.

The earliest statements that we have give the ratio as 3, later (1700 B.C.) as about 3.16, still later (220 B.C.) as $3\tfrac{1}{7}$. The most remarkable ratio of whole numbers that approximate to π (1000 A.D.) is $\tfrac{355}{113}$, which when expressed decimally agrees with π, expressed decimally, to the sixth decimal place.

$\tfrac{355}{113}$ may be easily remembered by noting the fact that if we write the first three odd numbers twice each, and then divide the last three by the first three, thus, 113)355, we have the ratio.

If 3.1415926 be expressed as a continual fraction, it gives:

$$\pi = 3 + \frac{1}{7} + \frac{1}{15} + \frac{1}{1} + \frac{1}{190} + \cdots$$

For which the first three succeeding convergents are:

$$C_1 = 3\tfrac{1}{7}\ = \tfrac{22}{7},$$
$$C_2 = 3\tfrac{15}{106} = \tfrac{333}{106},$$
$$C_3 = 3\tfrac{16}{113} = \tfrac{355}{113}.$$

The terms of the series, $\tfrac{1}{2}, \tfrac{1}{4}, \sqrt{\tfrac{1}{2}\cdot\tfrac{1}{4}}$, etc., are (beginning with the fourth term) alternately the arithmetical and the geometrical means of the two preceding terms; and is known as Schwab's series.

Remark. — Because of the fact that we are unable to express the area of a circle in terms of the square on the radius exactly, it does not follow that other areas bounded by curved lines, or by a combination of curved and straight lines, cannot be expressed exactly in terms of a square on *some* given line, or a rectangle, or a parallelogram, which may be converted into a square.

Illustration. — The locus of a point which moves in a plane so that the ratio of its distances from a fixed point and from a fixed line equals 1 generates what is called a *parabola*.

If through any point (P) a line (PP_1) parallel to the given line (MM_1) be drawn, the area included between the curve and PP_1 will be exactly equal to $\tfrac{2}{3}$ the rectangle PQQ_1P_1.

This fact will be established in § 163.

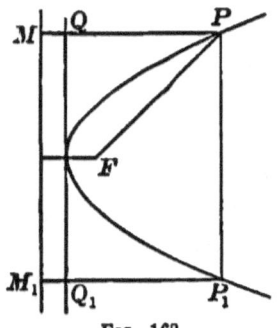

Fig. 163.

101. By Exs. 5 and 6, § 94, we see that the area of any regular polygon equals the rectangle on the half-perimeter and the apothem, or frequently expressed as half the product of perimeter and apothem. The same holds true no matter how many sides the polygon may have.

If successive polygons, doubling in number of sides, be inscribed within a circumference, the perimeter will continually increase, approaching the circumference of the circle as its limit, and the apothem will increase, approaching the radius of the circle as its limit.

If successive polygons, doubling in number of sides, be circumscribed about a circumference, the perimeter will continually decrease, but the apothem will remain constant.

In each case the area equals the half-product of perimeter and apothem.

When we have reached 32768 sides for each, the areas of the circumscribed and inscribed regular polygons (in terms of the square on the radius) agree to seven decimal places. The area of the circle lies between the two areas, and the circumference of the circle is the common limit upon which the perimeters of the inscribed and circumscribed polygons converge; we therefore say that the area of a circle equals half the product of its circumference and radius. If A represent the area of a circle, C its circumference, and R its radius,

$$A = \tfrac{1}{2} C \cdot R.$$

Expressed in general language:

The area of a circle equals the half product of the circumference and the radius.

102. From § 100,
$$A = \pi R^2. \quad \text{But } A = \tfrac{1}{2}CR.$$
$$\therefore \tfrac{1}{2}CR = \pi R^2, \text{ or } \tfrac{1}{2}C = \pi R.$$
$$C = 2\pi R, \qquad \text{(form most used)}$$
or $\qquad C = \pi D, \text{ or } \dfrac{C}{D} = \pi.$

Hence π may also be described as the ratio of the circumference to the diameter of a circle, as well as the ratio of the area of a circle to the square on its radius.

103. THEOREM. *The perimeter of* ANY *polygon inscribed within a circle is* LESS *than the circumference, and the perimeter of* ANY *polygon circumscribed about a circle is* GREATER *than the circumference of the circle.*

(*a*) Each side of the inscribed polygon is a straight line, joining two points, hence will be shorter than the arc which it subtends. The sum of these chords — which is the perimeter of the polygon — will therefore be less than the sum of the subtended arcs — which is the circumference of the circle.

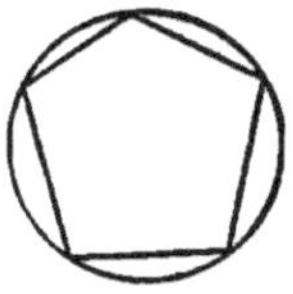

Fig. 169.

Q. E. D.

(*b*) By Ex. 6, § 94, the area of a circumscribed polygon equals the half-product of the perimeter and the radius of the circle.
$$A' = \tfrac{1}{2}PR.$$
By § 101, the area of the circle equals the half-product of its circumference and radius.
$$A = \tfrac{1}{2}CR.$$
But the area of the circumscribed polygon is greater than the area of the circle.
$$\therefore \tfrac{1}{2}PR > \tfrac{1}{2}CR, \text{ or } P > C. \qquad \text{Q. E. D.}$$

AREA OF A CIRCLE.

Exercises. — 1. Show that the area of any sector equals half the product of the included arc and the radius.

2. Find the areas of sectors of 30°, of 45°, of 120°, of 300°, of 390°, of 480°, and of 540°.

3. The areas of circles are to each other as the squares of their radii, or the squares of their diameters, or the squares of their circumferences, or the squares of any corresponding lines.

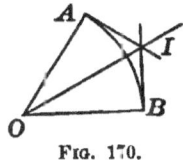

Fig. 170.

4. If at the extremities of an arc AB, which determines a sector of a circle, tangents be drawn intersecting at I, $AI + IB$ will be greater than the arc AB.

5. Show that if two circumferences intersect, the subtended arcs on each side of the common chord will be unequal, and the enveloping one will be the longer.

Show how to determine by tangents at A or B which curve will be the enveloping one. Discuss the problem thoroughly.

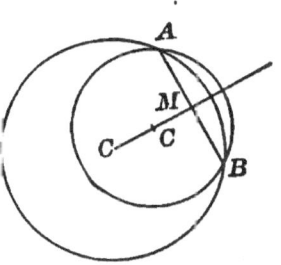

Fig. 171.

104. Problems. — 1. To inscribe a regular decagon in a circle.

Analysis. — If AB were the side of a regular inscribed decagon, and if we should join the vertices of the polygon to the centre, we should have ten triangles formed, each equal to ACB.

$$\angle ACB = 36°,$$
$$\angle CAB = 72°,$$
$$\angle CBA = 72°.$$

If the $\angle CBA$ be bisected by the line BD, we would have the $\triangle CBD$ and ABD isosceles as well as the $\triangle ACB$.

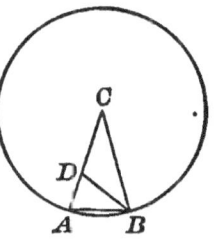

Fig. 172.

The $\triangle ACB$ and ABD being mutually equiangular, are similar.

$$\therefore \frac{CA}{AB} = \frac{BD}{DA}.$$

But
$$AB = BD = CD.$$
$$\therefore \frac{CA}{CD} = \frac{CD}{DA}.$$

Which means that the radius will be separated into mean and extreme ratio.

Construction. — By Prob. 13, § 91, separate the radius of the circle in which the decagon is to be inscribed, into parts such that the ratio of the whole radius to the larger part equals the ratio of the larger part to the smaller.
$$\frac{CK}{CQ} = \frac{CQ}{QK}.$$

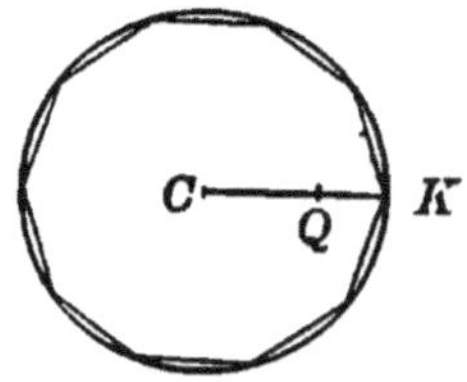

From K lay off chords equal to CQ; they will form the sides of the required decagon.

Fig. 173.

2. Show that if the alternate vertices of a decagon be joined, a regular pentagon will be formed.

3. Making use of § 88, show how to find the diagonals of a regular pentagon, having given the length of one side.

4. Show how a regular hexagon and a regular decagon may be used to inscribe a regular polygon of 15 sides, and hence one of 30 sides.

5. Show that the square on the side of a regular inscribed pentagon equals the sum of the squares on a side of the regular inscribed decagon and on the radius of the circle.

6. Having given the side of a regular decagon, find the corresponding radius.

7. Find the area of a regular inscribed dodecagon in terms of the square on the radius.

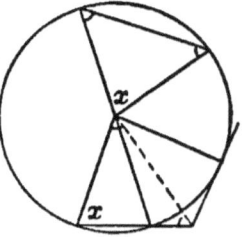

Fig. 174.

8. Show how to make a five-pointed star and determine the sum of the angles at the points.

SOME PROBLEMS IN PLANE GEOMETRY.

(NOTE. — Do not neglect the analysis.)

Show that:

1. The three altitudes of a triangle are concurrent lines.

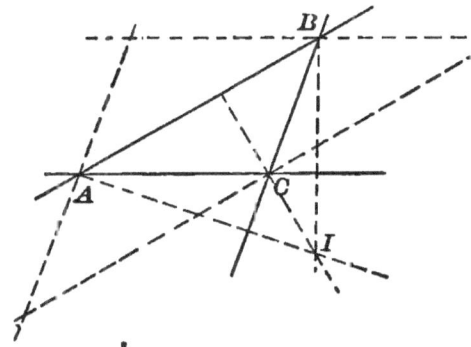

FIG. 175.

Suggestion. — Through the vertices of the given triangle draw lines parallel to the opposite sides.

2. A quadrangle, the opposite sides of which are equal, is a parallelogram.

3. Find the area of a triangle the sides of which are 12, 16, and 20. Find the length of the median from the largest angle. Determine the radii of the circumscribed and inscribed circles.

4. If from any point in the base of an isosceles triangle lines be drawn parallel to the equal sides, a parallelogram will be formed, the perimeter of which will equal the sum of the sides parallel, to which lines have been drawn.

5. If angle bisectors be drawn from the vertices of a scalene triangle, they will be unequal; and the greater one will be drawn through the vertex of the lesser angle.

6. The bisectors of the base angles of an isosceles triangle are equal.

7. The converse of the last exercise is true.

8. The number of sides of a polygon may be found when the sum of its interior angles is three times the sum of its exterior angles.

9. Show how to determine a line parallel to the parallels of a trapezoid that will bisect its area.

10. The lines joining the middle points of the adjacent sides of any quadrangle will form a parallelogram.

11. The lines joining the middle points of the adjacent sides of a rectangle will be a rhombus.

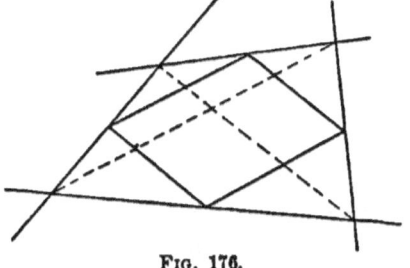

Fig. 176.

12. The lines joining the middle points of the adjacent sides of a rhombus will be a rectangle.

13. If two $\angle s$, A and B, of the $\triangle ABC$ be bisected, and a line DE be drawn through I, $\parallel$ to AB, $DE = AD + BE$.

14. The sum of the perpendiculars from any point in the base of an isosceles triangle to the equal sides

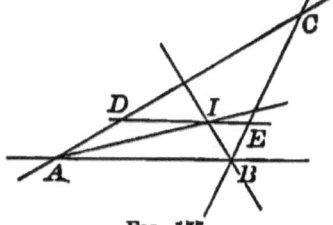

Fig. 177.

is constant, and equals the altitude from one of the equal angles.

Analysis. — If $DE + DF = AH$, and if we should through D draw a parallel to CB, it would determine $\overline{KH} = \overline{DF}$, and in order that the proposition be true, $\overline{AK}$ would have to equal $\overline{DE}$.

Proof. — $\overline{AK}$ does equal $\overline{DE}$, because the △ AKD and DEA are equal; being right triangles, having a common hypotenuse, and the $\angle KDA = \angle EAD$.

$\therefore \overline{AK} = \overline{DE}$, and $\overline{AK} + \overline{KH} = \overline{DE} + \overline{DF}$.   Q. E. D.

15. A median of a triangle is less than half the sum of the two sides which form the vertex from which the median is drawn.

16. The angles formed by the bisectors of two angles of a triangle will be such that one of them will equal 90° *plus* half the

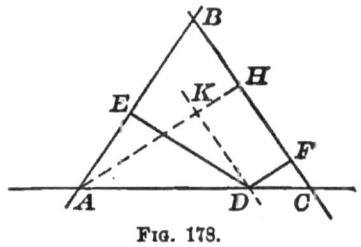

Fig. 178.

third angle of the triangle, and the other will equal 90° *minus* half the third angle of the triangle.

17. In any right triangle the median from the right angle equals half the hypotenuse.

18. If through the three vertices of a triangle the bisectors of the exterior angles be drawn, four new triangles will be formed, which will be mutually equiangular.

19. If the opposite angles of a quadrangle are equal, the figure will be a parallelogram.

20. A billiard ball shot parallel to one of the diagonals of a table, and making the angles of reflection equal to the angles of incidence, will return to the place from which it is shot.

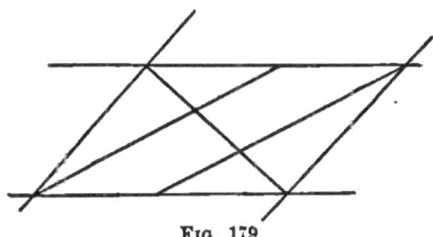

Fig. 179.

21. If a diagonal of a parallelogram be drawn, and lines be drawn from the middle points of a pair of opposite sides to the other vertices, they will trisect the diagonal.

22. The bisectors of the angles of a quadrangle form a second quadrangle, the opposite angles of which are supplementary.

SOME CONSTRUCTIONS.

(Do not neglect the analysis or the discussion.)

23. Find on a given line a point that shall be equally distant from two given points.

24. Find on a given circle points that shall be equally distant from two given points.

25. Trisect a right angle.

26. Trisect an angle of 72°.

NOTE. — The trisection of *any* angle is a construction that cannot be made by the geometry of the straight line and circle, but can be accomplished by the use of a number of curves, the treatment of which is beyond the scope of this book.

27. In a $\triangle ABC$ draw $DE \parallel$ to AC, so that $\overline{DE}$ shall equal $\overline{AD} + \overline{CE}$.

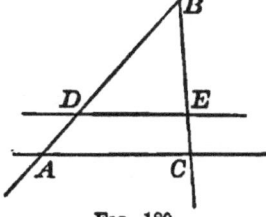

FIG. 180.

28. Construct an equilateral triangle, having given:
   (*a*) The altitude.
   (*b*) The perimeter.
   (*c*) The radius of the inscribed circle.
   (*d*) The radius of the circumscribed circle.

29. Construct an isosceles triangle, having given:
(*a*) The base and perimeter.
(*b*) The base and altitude.
(*c*) The base and angle at the vertex.
(*d*) The altitude and the perimeter.
(*e*) The altitude and the angle at the vertex.

30. Construct a triangle, having given:
(*a*) Two sides and their included median.
(*b*) An angle, a side opposite, and the difference of the adjacent sides.

(c) The base, the angle at the vertex, and the altitude.

(d) The base, the angle at the vertex, and the perpendicular from one of the base angles to the opposite side.

(e) The base, the angle at the vertex, and the median to the base.

(f) The angles and the perimeter.

(g) Two sides and the altitude to the vertex they form.

(h) Two sides and the altitude on one of them.

(i) An angle, the side opposite, and the sum of the other two sides.

(j) An angle of 90°, and the radii of the inscribed and circumscribed circles.

(k) The centres of the escribed circles.

(l) The three altitudes.

31. At the extremity of a given straight line segment to construct a perpendicular without producing the segment.

32. Through a given point within a circle to draw the shortest chord.

33. Find the centre of a circle of given radius that shall pass through two given points.

34. In a given circle draw a concentric circle, so that the diameter of the inner circle shall equal a chord of the outer circle drawn tangent to the inner one.

35. In three different ways show how to construct a segment of a circle that will contain a given angle.

36. Let A, B, and C represent three fixed points on shore; S, the point at which a sounding is made by one occupant of a boat, while another takes the angles subtended by $\overline{AB}$ and $\overline{BC}$.

Show how the point S may be located on a map which gives the positions of A, B, and C.

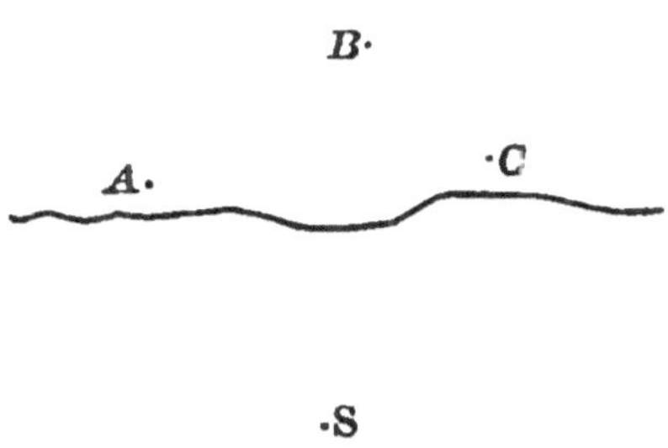

Fig. 131.

37. Through a given point not in a given straight line to draw a line that shall make a given angle with the given line.

38. To a given circle draw a tangent that shall be parallel to a given straight line.

39. In a given circle to draw a chord of given length that shall be parallel to a given line.

40. Through a given point within a circle to draw a chord of given length.

41. Construct a circle of given radius that shall be tangent to a given circle, and also to a given straight line.

42. Construct a circle that shall cut three equal chords from three given straight lines.

How will it be if these chords are to be of given length?

43. Construct the point through which the bisector of the vertical angle of a triangle, having a fixed base, will pass.

44. With the vertices of a triangle as centres, construct three circles so that each shall be tangent to the other two.

45. Construct a rectangle, having given a diagonal and one side.

46. Will the sum of diagonal and radius of the inscribed circle be sufficient to determine a square?

47. Construct a square having given the sum of the radii of the circumscribed and inscribed circles.

48. Construct a rectangle, having given its perimeter and a diagonal.

49. Construct as many circles as possible of given radius that shall be tangent to a given circle of the same radius and tangent to each other.

THEOREMS.

50. If ABC be an inscribed equilateral triangle, and P be any point on the $\overparen{AB}$,

$$\overline{PC} = \overline{PA} + \overline{PB}.$$

51. The distance of each side of the $\triangle ABC$ (Ex. 50) from the centre of the circle is half the radius of the circle.

52. If secants be drawn through the points of intersection of two circles, the chords joining the other points in which the secants intersect the circles will be parallel.

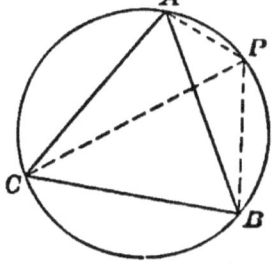

Fig. 182.

PROBLEMS.   143

Query.— How will it be if the circles are tangent?

53. If diameters be drawn to one of the points of intersection of two circles, and the other extremities be joined by a straight line, that line will pass through the other point of intersection of the circles.

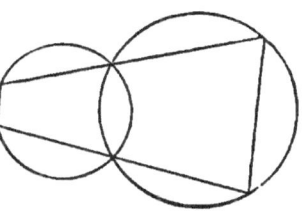

FIG. 183.

54. If PA and PB be two tangents to a circle from an exterior point P, M be a moving point in the circumference, and QH be a

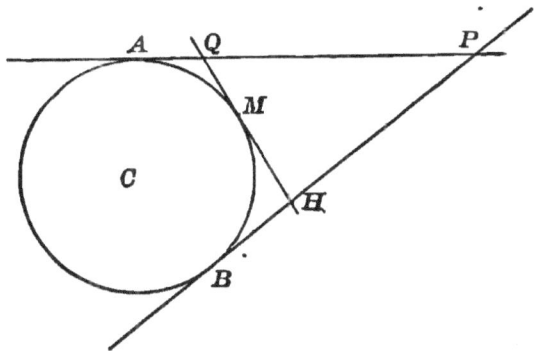

FIG. 184.

tangent at M, the perimeter of the $\triangle PQH$ will be constant and $= \overline{PA} + \overline{PB}$.

Remark. — In the discussion do not limit M to the smaller $\widehat{AB}$. Look for positive and negative distances.

55. The angle at C (Ex. 54) subtended by $\overline{QH}$ will be constant and $< 90°$ when M is on the smaller $\widehat{AB}$; it will also be constant, but $> 90°$, when M is on the larger $\widehat{AB}$.

56. If in a circle two equal chords intersect, the segments of one will be equal to the segments of the other.

57. If the angles P and Q of the quadrilateral $ABPCQD$ be bisected by PI and QI,
$\angle I = \angle A + \angle a + \angle \beta$;
also,
$\angle I = \dfrac{\angle A + \angle C}{2}$.

If $\angle A + \angle C = 180°$,
$\angle I = 90°$.

Remark.—In the latter case a circle could be circumscribed about the quadrangle $ABCD$.

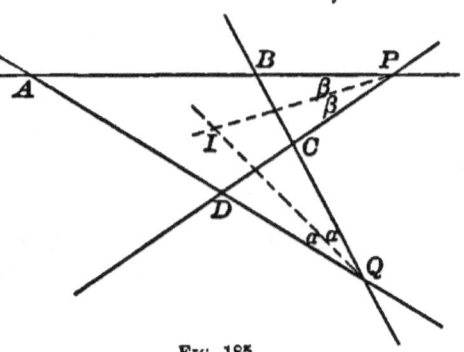

Fig. 185.

58. The portions of any straight line included between the circumferences of concentric circles will be equal.

59. If a quadrangle be circumscribed about a circle, the sum of a pair of non-adjacent sides will equal the sum of the other pair.

60. The converse of Ex. 59 is true.

61. The common secants of three intersecting circles are concurrent.

See Exs. 10 and 11, § 91.

62. The three circles which pass through the vertices of a triangle, and intersect two and two on the sides, will all pass through a common point.

Establish the converse.

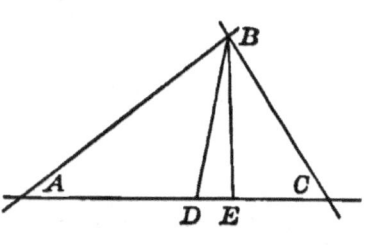

Fig. 187.

Fig. 186.

63. If $\overline{BD}$ bisects $\angle ABC$, and $\overline{BE}$ is a $\perp$ to AC, the

$$\angle DBE = \frac{C-A}{2}.$$

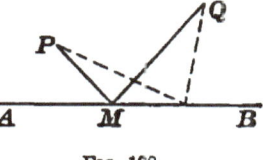

64. If from two points, P and Q, lines be drawn to a moving point M in the line AB, the sum of $\overline{PM}$ and $\overline{QM}$ will be a minimum, when they make equal angles with AB.

FIG. 188.

Query. — How will it be if the points are on opposite sides of AB?

SOME PROBLEMS IN LOCI.

NOTE. — In general, to attack a problem in loci, make constructions enough under the given conditions to *suggest* a locus. Then follows the analysis and the demonstration: *If* the *suggested* locus be the required one, any (every) position on this suggested locus will fulfil the required conditions; and any (every) position not on the suggested locus will *not* fulfil the required conditions.

65. Find the locus of the middle point of a given straight line segment, the extremities of which remain in two lines at right angles to each other.

Let the lines OY and OX be at right angles to each other; and let S be the given segment, of which M is the middle point.

Seven or eight constructions under the conditions of the problem are enough to suggest that the locus is

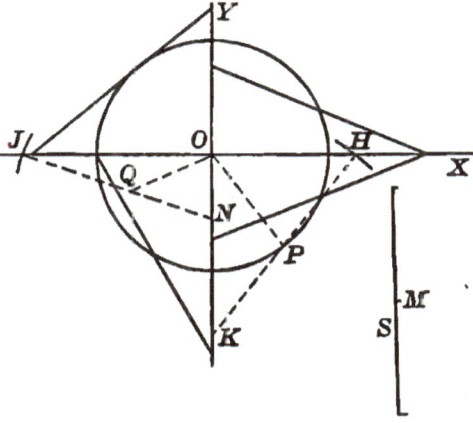

FIG. 189.

the circumference of a circle, with its centre at O, and its radius equal to half the given segment.

L

If the *suggested* locus *be* the required one, any point, as P, on the circumference will fulfil the required conditions; and *any* point, as Q, not on the circumference, will not fulfil the required conditions.

With P as a centre and half of S as a radius construct an arc intersecting OX at H. Join P and H; draw the auxiliary segment OP; and produce HP until it intersects OY at K.

The $\triangle OPH$ is isosceles

(two sides being equal).

The $\triangle OPK$ is isosceles

(two angles being equal).

$\therefore PH = PO = PK$.

$\therefore HK = S$ and is bisected at P.

To establish the fact that Q will not fulfil the required conditions proceed in the same way:

With Q as a centre and half of S as a radius, construct an arc intersecting OX at J. Draw the auxiliary OQ.

$QO < QJ$.

$\therefore \angle QJO < \angle QOJ$.

$\therefore \angle QNO > \angle QON$.

$\therefore \overline{QN} < \overline{QO} < \overline{QJ}$.

That is: a segment of length S cannot be drawn through Q, so as to be bisected at Q, and have its extremities in the given lines.

The student will show, after the same manner, that a point without the circle will not fulfil the required conditions; and so cannot be a point on the locus.

NOTE.—Any other point of the line of given length will generate a curve, but that curve will be an ellipse, the shape of which will depend upon the point selected.

66. Find the locus of the middle points of the chords of given length in a given circle.

67. Find the locus of the point P, as M moves about the circumference; $\overline{PM}$ remaining constant in length, and parallel to its initial position.

68. Find the locus of the middle points of all straight line segments included between two given parallels.

69. Find the locus of the middle points of all chords that can be drawn through a given point in a circle.

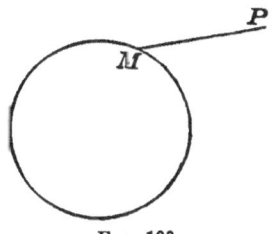

Fig. 190.

70. Find the locus of the middle points of all secants through a given point to a given circle.

71. Find the locus of the centre of a circle of given radius that rolls on a straight line.

72. Find the locus of the centre of a circle of given radius that rolls on the circumference of a given circle.

73. AB is a diameter of a circle; through A a secant is drawn; at T, a tangent, and through B a secant perpendicular to the tangent. Find the locus of the intersection of the two secants.

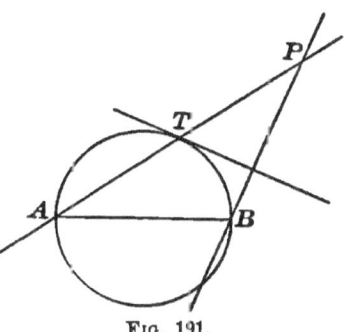

Fig. 191.

74. Let θ change from 0° to 360°, and $CP = NH$. Find the locus of P.

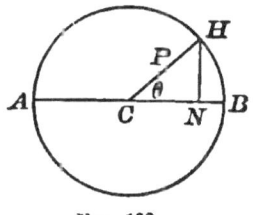

Fig. 192.

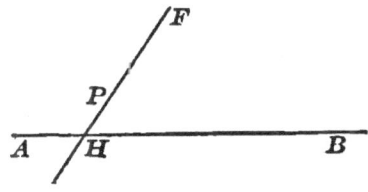

Fig. 193.

75. AB is a fixed line; and F a given point. Locate P, so that $\dfrac{FP}{PH} = \dfrac{m}{n}$ (constant).

76. AB and CD are given lines. Find the locus of P, which moves so that the sum of the perpendiculars on the given lines equals q.

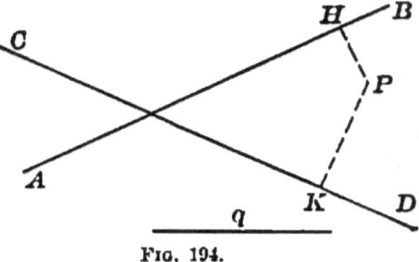

Fig. 194.

NOTE. — When perpendiculars to a given line are being considered, the distances on one side are positive, and on the other side negative. This should be borne in mind in 74 and 76.

77. AB and CD are fixed at right angles to each other. Find the locus of P, $\overline{PH}^2 + \overline{PK}^2$ being constant.

78. A triangle has a given base and vertical angle. Find the locus of its vertex.

79. Find the locus of the centre of the inscribed circle; the triangle being as described in Ex. 78.

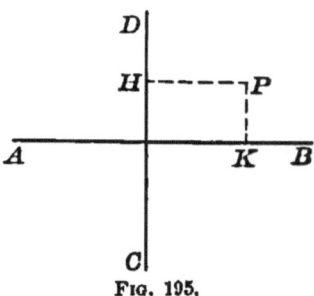

Fig. 195.

80. In the same triangle, find the locus of the point of intersection of the altitudes.

81. Find the locus of the centre of a circle, tangent to two given straight lines.

Remark. — Do not fail in each problem to discuss the particular cases.

82. Find the locus of a point, the sum of the distances of which from two vertices of an equilateral triangle equals its distance from the third vertex.

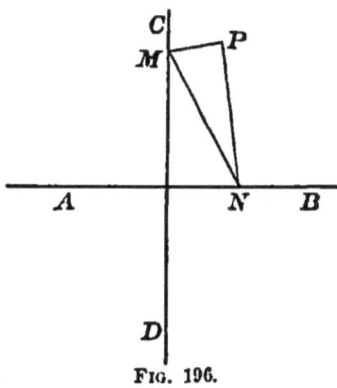

Fig. 196.

83. Find the locus of points from which a given circle will subtend a given angle.

PROBLEMS. 149

84. MNP is a right triangle, the oblique vertices of which remain in AB and CD, that are fixed at right angles to each other. Find the locus of P.

85. $\overline{PA}^2 + \overline{PB}^2 = \overline{AB}^2$. Find the locus of P.

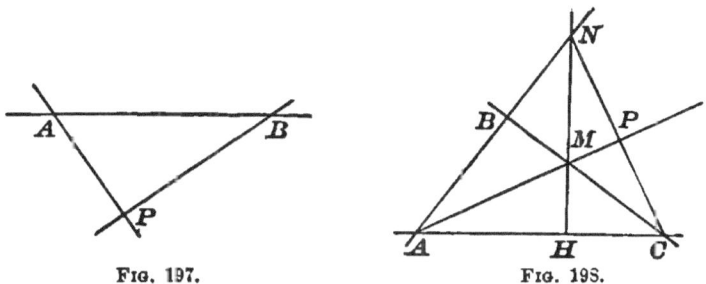

FIG. 197.   FIG. 198.

86. ABC is a given right triangle; HN is a moving perpendicular; AM and CN meet at P. Find the locus of P.

SOME PROBLEMS IN CONSTRUCTION.

87. Through a given point within an angle to draw a line so that the segments included between the lines shall be bisected at the point.

Let IR and IQ represent the lines forming the angle, and G the given point.

Analysis.— If HK were the required line (without saying whether it is or is not), and if through G an auxiliary line were drawn ∥ to IH, it would bisect IK; or *if* the auxiliary line were drawn ∥ to IK, it would bisect IH. Hence the

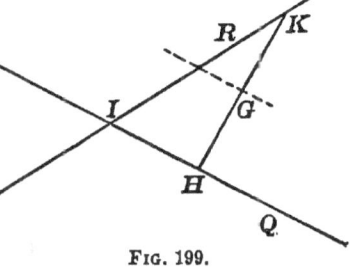

FIG. 199.

Construction.— Through the given point draw an auxiliary line parallel to one of the lines forming the angle. Let its intersection with the other one be A. Lay off $AH = IA$, and draw the line HG. The segment HK will be bisected at G as required.

Discussion. — If G should lie in an angle bisector, the $\triangle KIH$ would be isosceles. If G should lie in one of the lines forming the angle, $\overline{HK}$ would coincide with that line. If G should coincide with I, there would not be any construction.

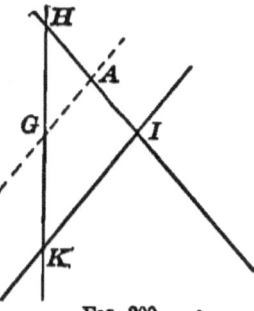

Note. — Exercise 87 has thus been given in detail, not because of its difficulty, but because it is desirable to present to the student the method of investigation that is to be pursued in all problems of construction, whether they be easy or difficult.

Fig. 200.

While in general, in the text-book solution of a problem, but one figure is used for the Analysis and the Proof, it is advisable that the student should use two.

88. AB and CD are fixed at right angles to each other, and EF is any other line. To construct a square such that two sides shall lie on AB and CD, and one vertex shall lie on EF.

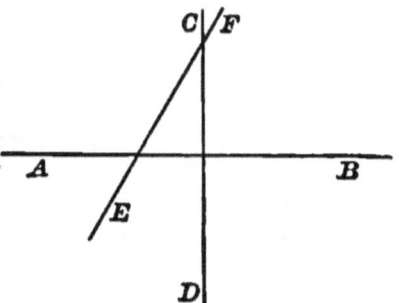

89. Construct a square, having given:

(*a*) The diagonal.

(*b*) The sum of a diagonal and one side.

(*c*) The difference of a diagonal and one side.

Fig. 201.

90. Construct a rectangle, having given:

(*a*) A diagonal and the difference of two adjacent sides.

(*b*) One side and the corresponding angle formed by the diagonals.

(*c*) One side and the sum of the diagonals.

91. Construct a parallelogram, having given:

(*a*) One side and the diagonals.

(*b*) The diagonals and one angle of the parallelogram.

(*c*) One side, one diagonal, and one angle.

92. Construct a circle of given radius, that shall intersect a given circle, so that they shall be at right angles to each other at a given point.

93. Construct a circle with a given centre which shall intersect a given circle at right angles.

94. Construct a circle of given radius :

(*a*) That shall be tangent to two given straight lines.

(*b*) That shall be tangent to a given line and to a given circle.

(*c*) That shall be tangent to two given circles.

(*d*) That shall pass through a given point, and be tangent to a given line.

(*e*) That shall pass through a given point, and be tangent to a given circle.

95. Construct a circle which shall be tangent :

(*a*) To two given lines, and pass through a given point.

(*b*) To a given line at a given point, and that shall pass through another given point.

(*c*) To a given circle at a given point, and tangent to a given straight line.

(*d*) To a given circle at a given point, and shall pass through another given point.

(*e*) To a given line at a given point, and also tangent to a given circle.

Let *C* represent the centre of the given circle, *AG* the given line, and *G* the point at which the required circle is to be tangent.

Analysis. — *If* the circle, the centre of which is *H*, *were* the required circle, and *if* at *G* a perpendicular *were* erected it would pass through *H*. *CT* would also pass through *H*, and *TG* would form with *TH* and *GH* an isosceles triangle. Then *if* through *C* a parallel to *TG*

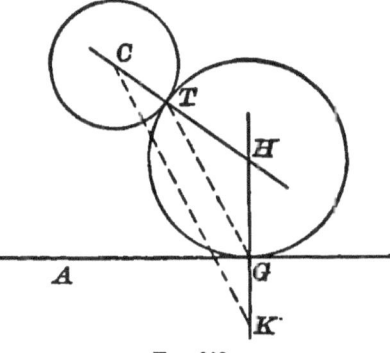

FIG. 202.

were drawn, it would form an isosceles △ CHK, and GK would equal CT.

Construction. — At the given point of the given line erect a perpendicular; on the opposite side of the line from the given circle, lay off $\overline{GK}$, the radius of the given circle. Draw CK. At C construct an ∠ KCH = ∠ CKG. It will determine H, the centre of the required circle.

With H as a centre and a radius HG, construct a circle; it will fulfil the required conditions. (*H* might have been as well determined by a perpendicular erected at the middle point of CK; or T might have been determined by drawing through G a parallel to KC.)

Discussion. — If a circle tangent to AG at G should increase in radius, beginning with a radius (0), it would, when the radius equals HG, be tangent to the given circle. Beyond that it would be an intersecting circle for a time, and then in one position be tangent to the given circle. Beyond that it would be an enveloping circle.

Let us now see about the second solution that has been suggested by the discussion.

Analysis. — *If* the circle the centre of which is H' were an enveloping tangent circle, and also a tangent to AG at G; a perpendicular erected at G would pass through H'; $T'C$ would pass through H'; the △ $T'H'G$ would be isosceles; and CK' ∥ to $T'G$ would determine $GK' = CT'$. Hence the

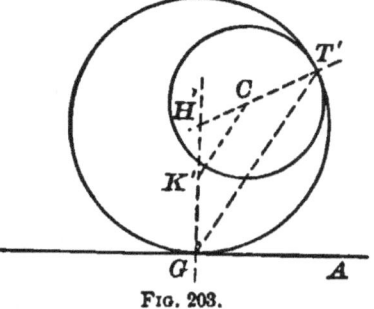

FIG. 203.

Construction. — Erect a perpendicular at G. Lay off GK', on the side *toward* the given circle, equal to the radius of the given circle. Draw CK'. Through G draw a parallel to CK', thus determining T'. The intersection of $T'C$ with the perpendicular at G, determines H', and $H'G$ is the radius of the required circle.

With H' as a centre, and a radius $\overline{H'G}$ construct the required circle.

PROBLEMS. 153

Further Discussion. — If AG be tangent to the given circle at Q, there will be but one construction.

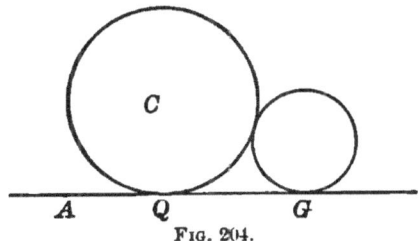
FIG. 204.

If G and Q coincide, there will be an infinite number of solutions.

If AG intersects the given circle, and G is without, there will be two constructions.

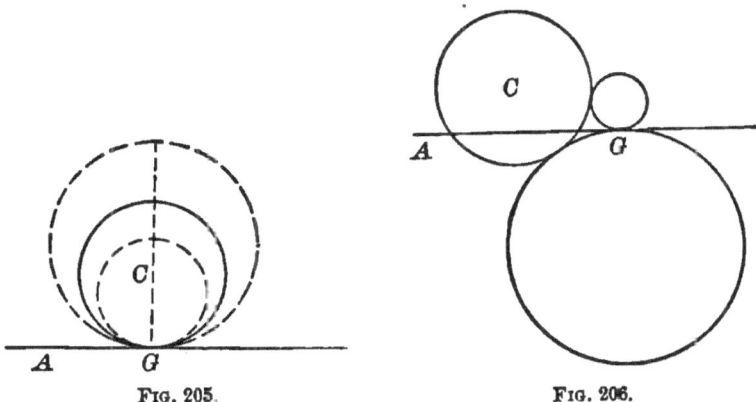
FIG. 205.   FIG. 206.

If AG intersects the given circle, and G is on the circumference, there will not be any solution.

If G should lie within the given circle, there will be two constructions.

In order to reach these different constructions, AG may be conceived as moving from one position to another that is parallel; and the point G may be conceived as moving in the line AG. It is interesting to note the special cases; *i.e.* the point and the straight line toward which the circles approach as *G* approaches some of its particular positions.

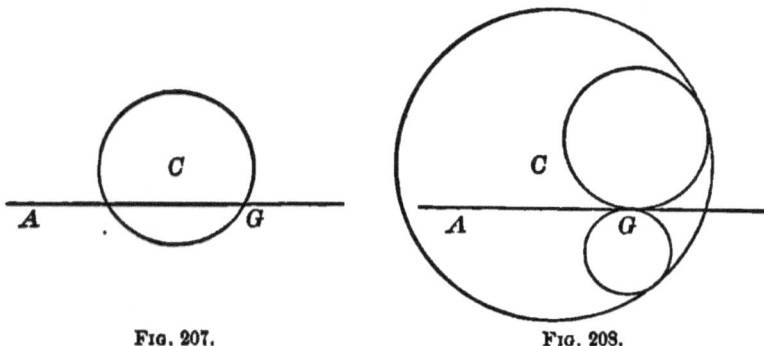

Fig. 207.   Fig. 208.

96. Through a point to draw a line that shall meet two given lines, which intersect outside the limits of the drawing. In practice on land it would read: "Which intersect in some inaccessible point."

97. Construct a circle that shall be tangent to a given line at a given point, and shall pass through a given point.

98. Pass the circumference of a circle through two points, and tangent to a given straight line.

99. Construct a triangle, having given one angle, the radius of the inscribed circle, and the radius of the corresponding escribed circle.

100. Construct a triangle, having given one vertex, and the feet of two altitudes.

101. Construct a triangle, having given the feet of the altitudes.

102. Use some of the following figures to establish the fact that the square on the hypothenuse of a right triangle equals the sum of the squares on the other sides.

NOTE.—These figures have come from various sources, principally, however, from students who have been called upon for original demonstrations. The number might have been largely increased, but a large enough number has been presented to show that there may be more than *one* way of solving a problem.

PROBLEMS.

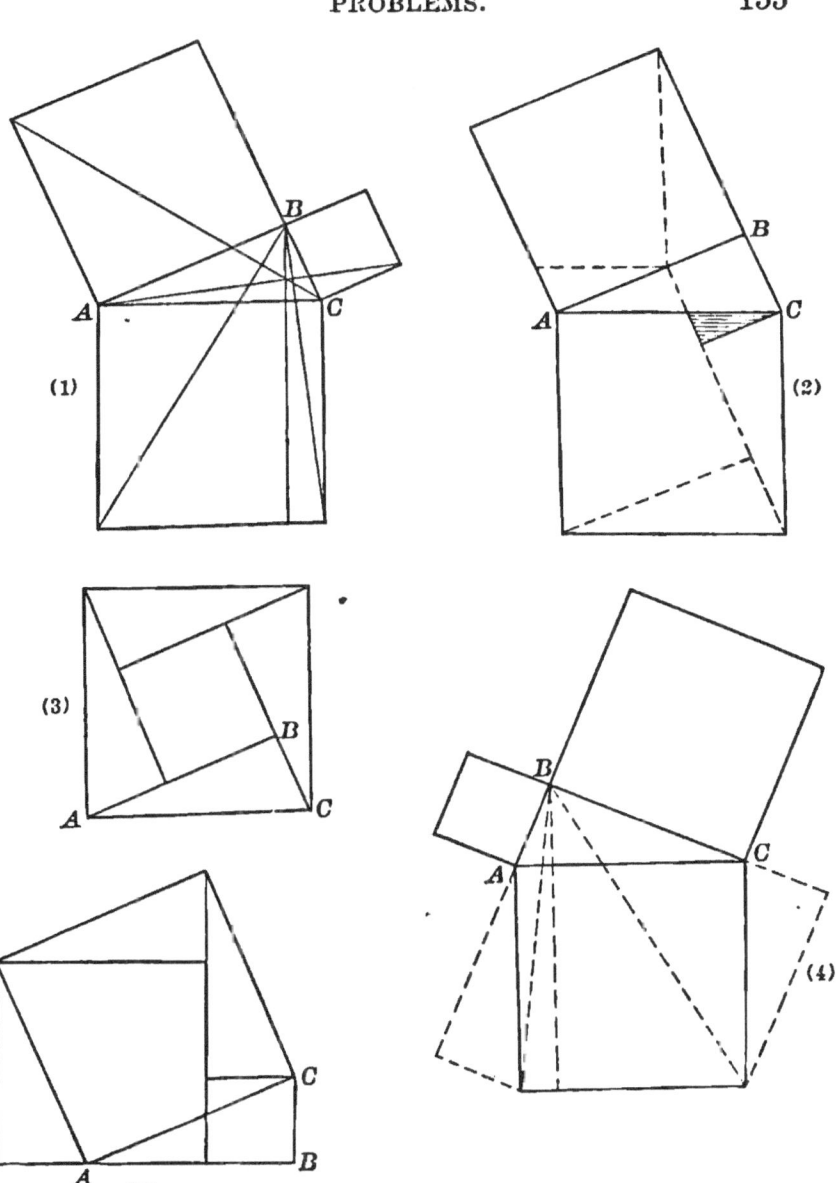

Fig. 209.

156 ELEMENTS OF GEOMETRY.

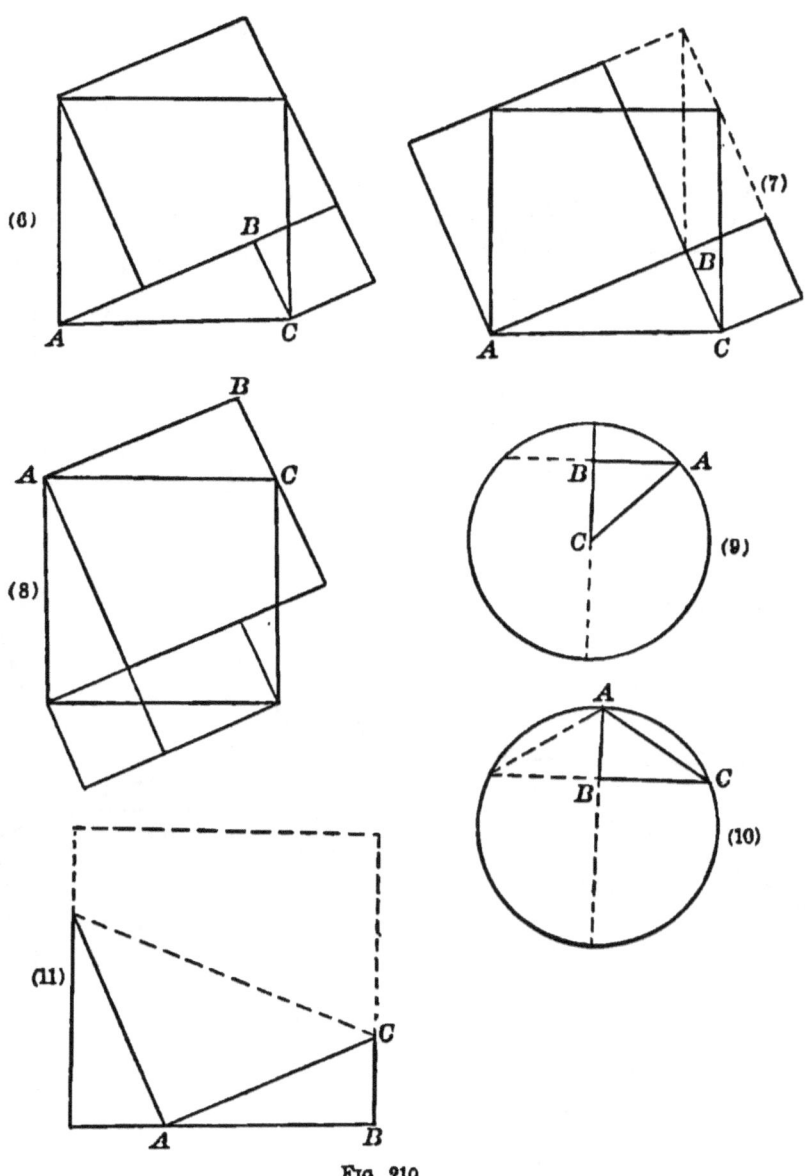

Fig. 210.

PROBLEMS. 157

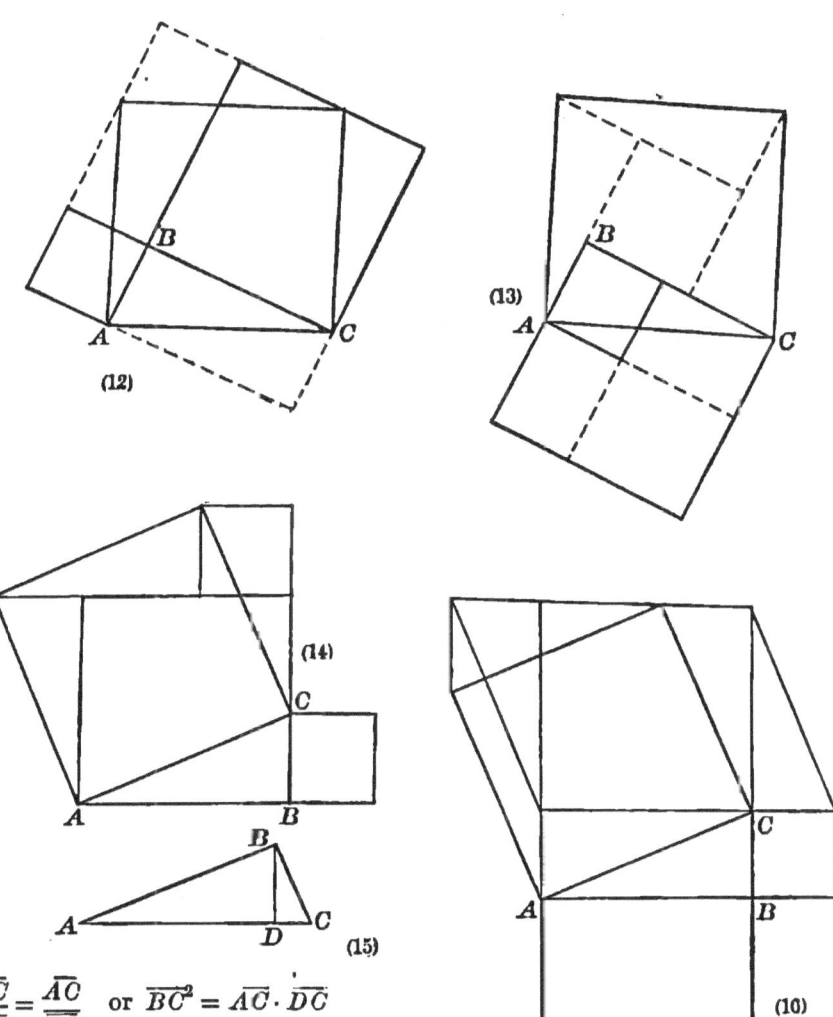

$\dfrac{\overline{BC}}{\overline{DC}} = \dfrac{\overline{AC}}{\overline{BC}}$ or $\overline{BC}^2 = \overline{AC} \cdot \overline{DC}$

$\dfrac{\overline{AB}}{\overline{AD}} = \dfrac{AC}{AB}$ or $\overline{AB}^2 = \overline{AC} \cdot \overline{AD}$

$\overline{BC}^2 + \overline{AB}^2 = \overline{AC}(\overline{AD} + \overline{DC})$

Fig. 211.

158   ELEMENTS OF GEOMETRY.

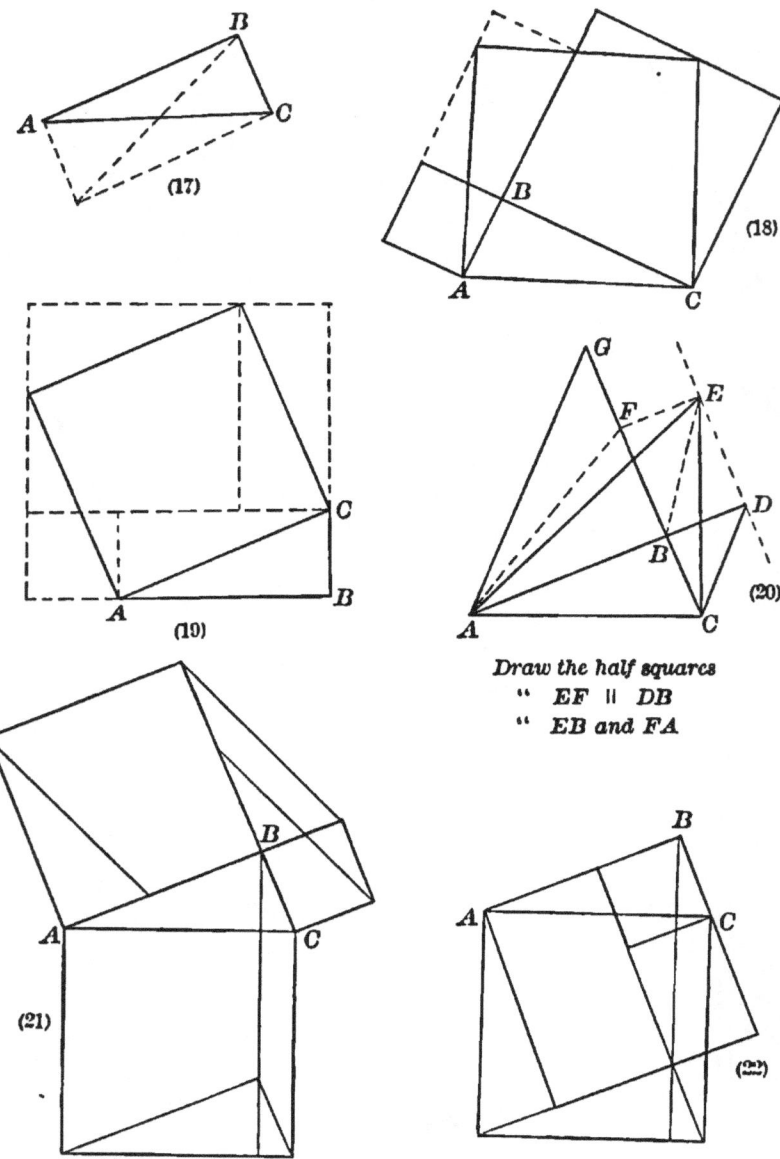

Draw the half squares
" EF ∥ DB
" EB and FA

Fig. 212.

PROBLEMS.

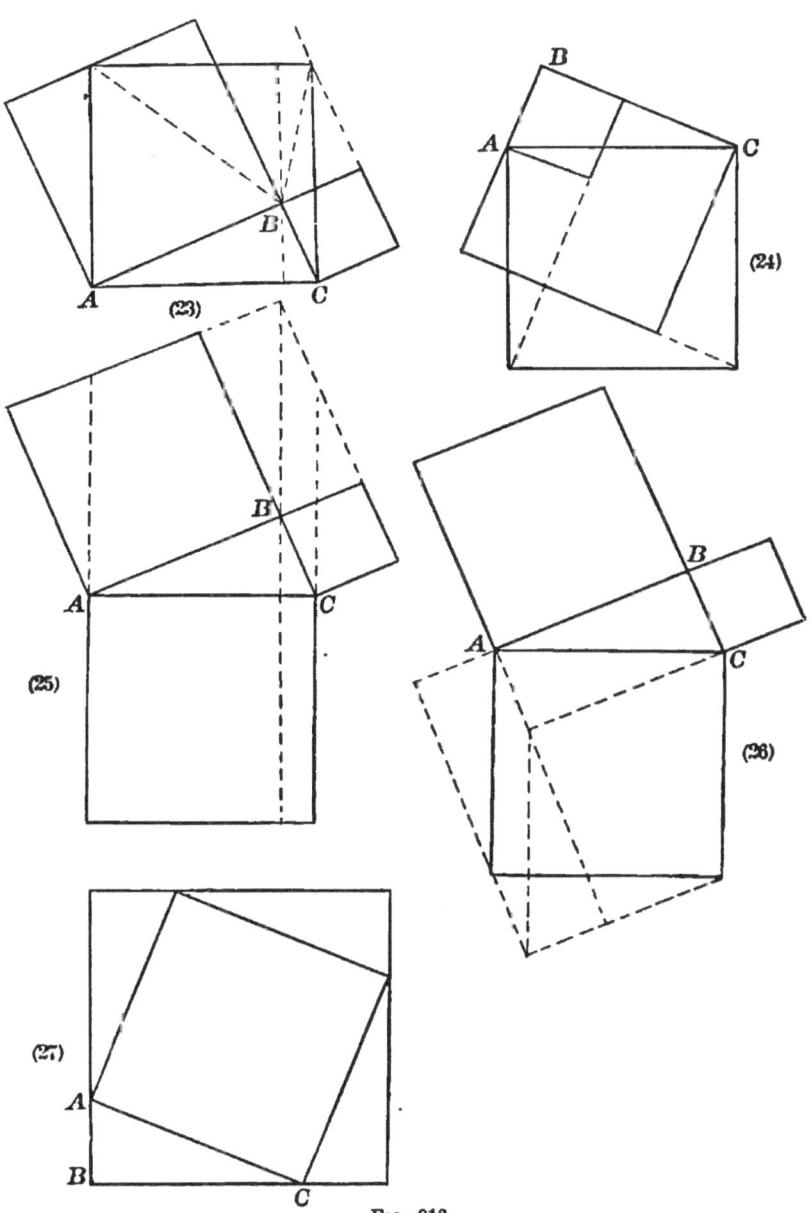

Fig. 213.

160   ELEMENTS OF GEOMETRY.

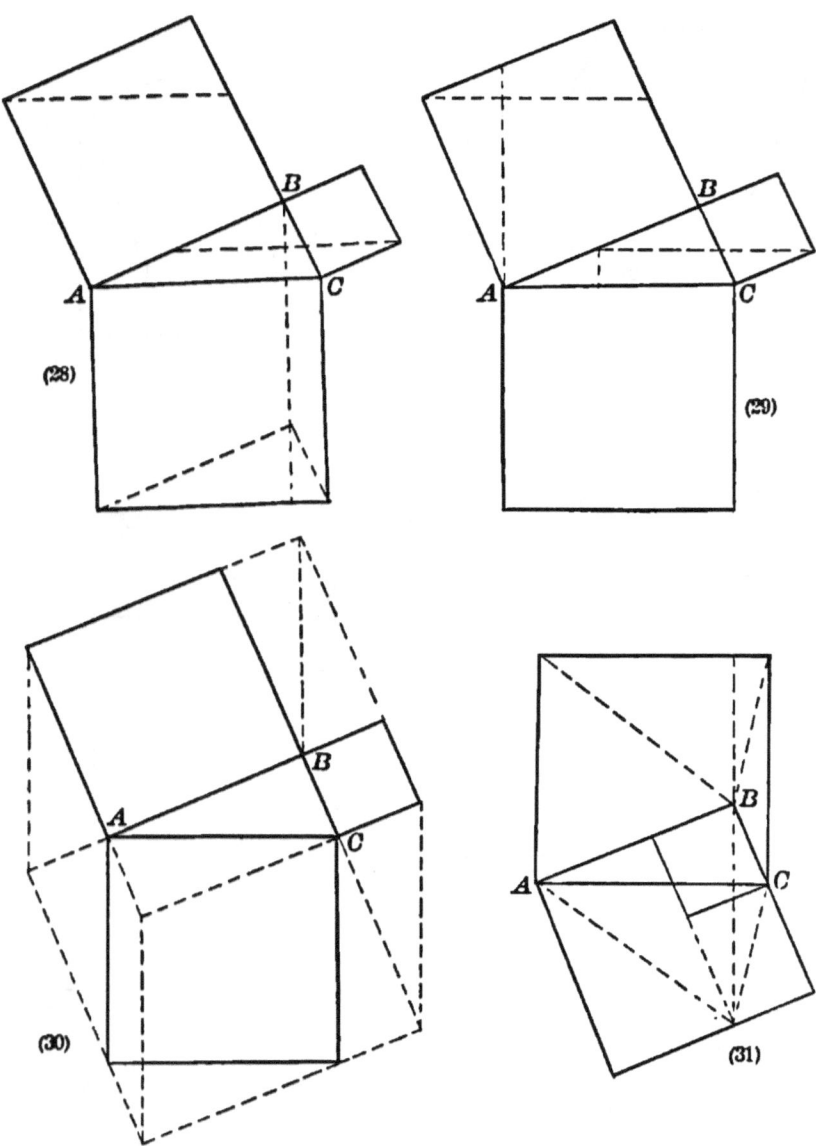

Fig. 214.

PROBLEMS. 161

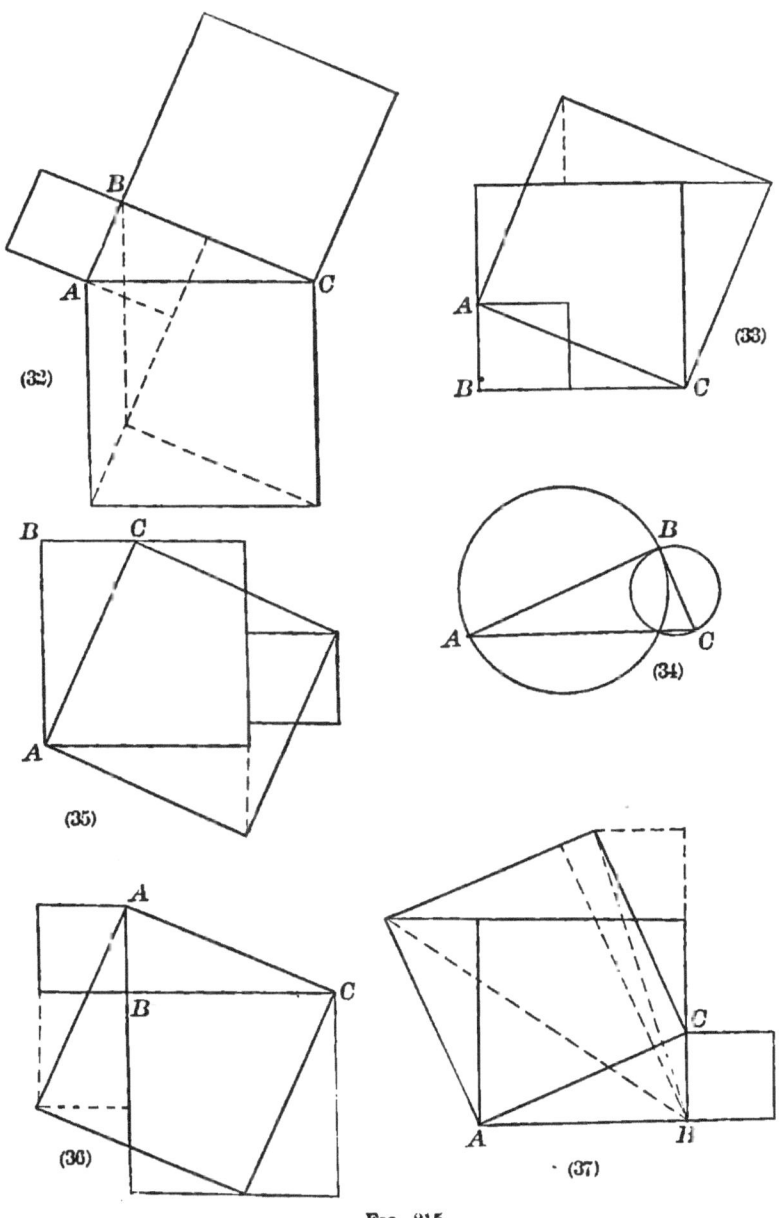

Fig. 215.

162 ELEMENTS OF GEOMETRY.

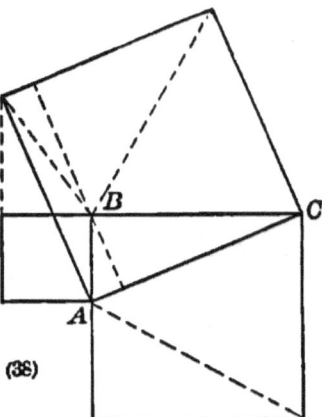

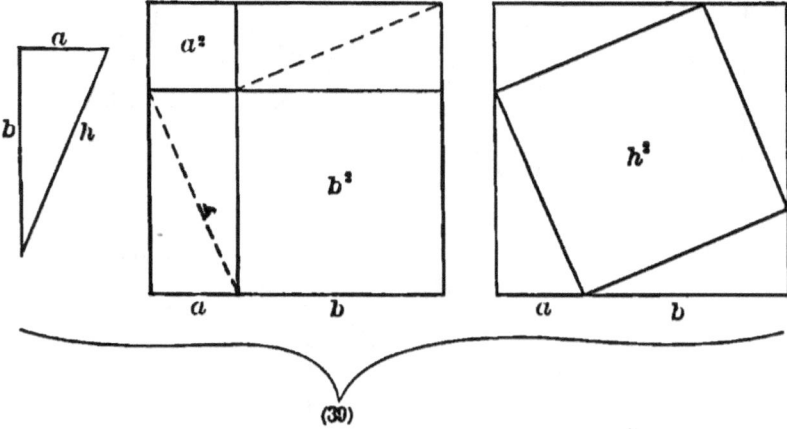

Fig. 216.

The most noted figures are: (1), (3), (5), (6), (25), and (27).
The simplest demonstrations are made by the figure already used in the body of the text, and by (5), (6), (24), and (39).

PROBLEMS.

103. How many trees can be planted on 160 acres of land:

(*a*) When set out in equilateral triangles, each side being 20 feet?

NOTE. — This manner of placing trees is sometimes called the hexagonal method; as any given tree (not on the outer boundary) will have six trees at the same distance from it, forming the vertices of a hexagon, of which the given tree is the centre.

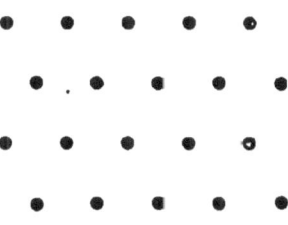

(*b*) When set out in squares, each side being 20 feet?

(*c*) When planted quincunx, 20 feet apart? *

104. What regular polygons may be used by themselves to cover a plane surface?

What combinations of regular polygons may be used for the same purpose?

105. Why does the honey bee build a hexagonal cell?

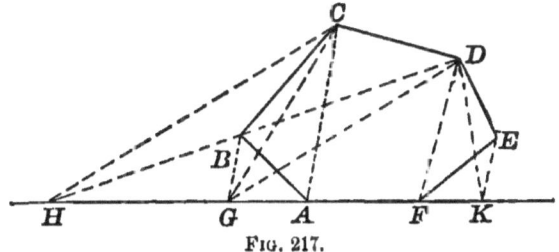

FIG. 217.

106. To convert a polygon into an equivalent triangle.

(*a*) If the given polygon be convex, and it be desired to have one side of the triangle in the line of one side of the polygon, as AF:

---

* NOTE. — Any tree (not on the outer boundary) will be the centre of a square, each vertex of which will be 20 feet from the centre.

Cut off the △ ABC, and put on its equivalent △ AGC.

Cut off the △ GCD, and put on its equivalent △ GHD.

Cut off the △ FED, and put on its equivaient △ FKD.

The polygon is reduced.

(b) If the given polygon be re-entrant, put on and cut off, until it is changed to a convex polygon, and then proceed as before.

107. Show that the diagonals of a trapezoid separate it into four triangles, two of which are similar, and the other two are equivalent in area.

108. Depending on the properties of similar triangles, show how to determine the direction in which to run a tunnel so that, starting at a given point it shall run in a straight line under a mountain, and shall emerge from the surface at another given point.

109. Show geometrically how to find the distance across a river in 6 different ways.

Do not neglect the simplest one.

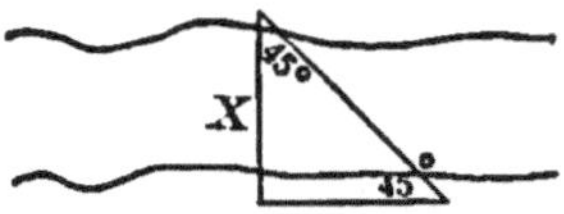

FIG. 218.

110. Construct a scale that shall measure hundredths.

FIG. 219.

111. Construct a vernier that shall measure hundredths.

Let the space from A to B represent one of the divisions of a scale which is subdivided as indicated in the drawing into 10 equal parts (if we use the decimal subdivision).

PROBLEMS. 165

The vernier slides along on the scale and has the distance $\overline{PQ}$ (which in length equals 9 of the subdivisions on the scale), separated into 10 equal parts. Then each subdivision on the vernier will be less by *one-tenth* than each subdivision on the scale.

Let us read the distance of the point P, from the 0 (zero) end of the scale. It is 3.18, an approximation, of course, but the error is less than .01. The student will supply the reasons for the reading.

NOTE.— Verniers are in great variety, and are called into use in cases where accuracy of observation is required. The levelling rods of the surveyor read to thousandths of a foot, and some of the circles used by surveyors in the measurement of angles read to ten seconds.

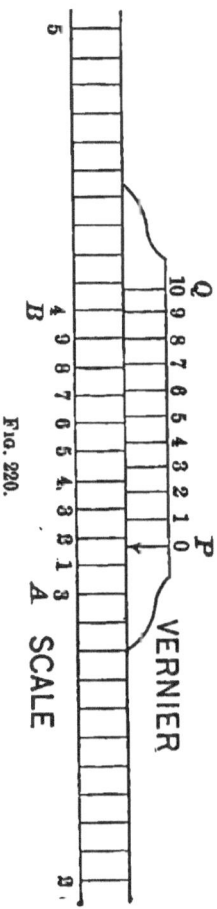

FIG. 220.

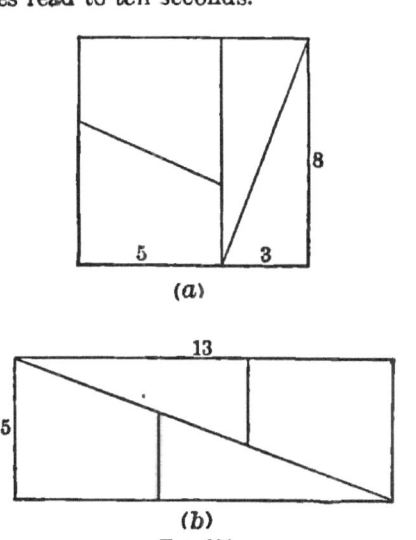

FIG. 221.

112. If a piece of paper 8 units square be cut as indicated in (*a*), and then be placed as in (*b*), 64 square units apparently become 65 square units. Where is the deception?

113. Construct a square that shall have one side on the base of a triangle, and its other vertices in the other two sides of the triangle.

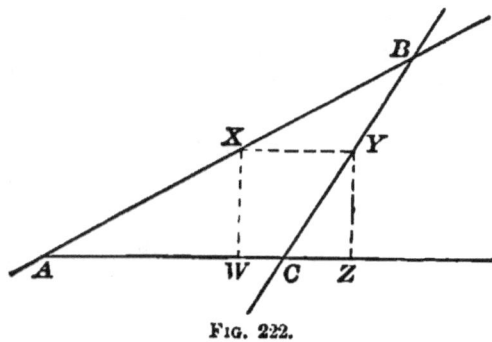

Fig. 222.

114. If straight lines be drawn from the vertices through any point within a triangle,

$$\frac{\overline{PI}}{\overline{PB}} + \frac{\overline{QI}}{\overline{QC}} + \frac{\overline{KI}}{\overline{KA}} = 1.$$

115. If from any point within a triangle, straight lines be drawn to the sides, and through the vertices lines be drawn parallel to them, then will

$$\frac{\overline{DP}}{\overline{HB}} + \frac{\overline{EP}}{\overline{KC}} + \frac{\overline{FP}}{\overline{MA}} = 1.$$

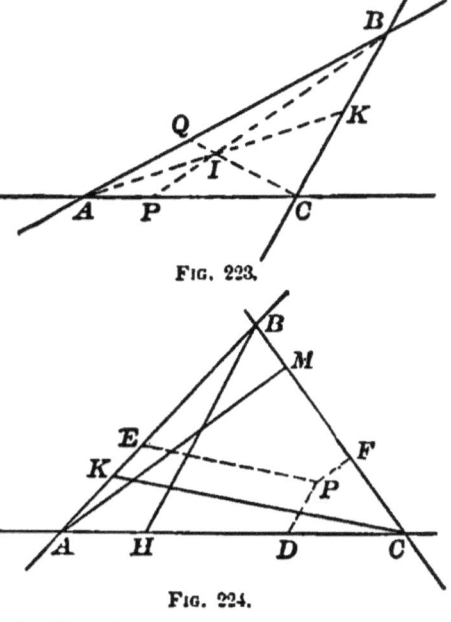

Fig. 223.

116. (*a*) Show that a point may be located between the extremities of a given segment, so that the ratio of its distances from the extremities shall equal *any* assumed ratio.

Fig. 224.

(*b*) Show that a point on the line, but exterior to the segment, may be found which will fulfil the same conditions.

PROBLEMS. 167

117. Approximately, what will be the relative amounts of water, that under the same head will be discharged from an inch pipe and from a quarter inch pipe.

118. Having given one side of a decagon 10, find the radius.

119. One side of a pentagon is 10; find a diagonal.

120. The radius of a circle being 1, find the area of an inscribed regular:

    (*a*) Triangle.   (*e*) Octagon.
    (*b*) Quadrangle.   (*f*) Decagon.
    (*c*) Pentagon.   (*g*) Dodecagon.
    (*d*) Hexagon.

121. The radius of the circumscribed circle being 10, find the side and the apothem of the regular inscribed decagon.

122. Compare the areas of the following regular figures, which have the same perimeter:

    (*a*) Triangle.   (*e*) Octagon.
    (*b*) Quadrangle.   (*f*) Decagon.
    (*c*) Pentagon.   (*g*) Dodecagon.
    (*d*) Hexagon.   (*h*) Circle.

123. Three circumferences, each of radius r, are tangent to each other. Find the included area.

124. The front and rear wheels of a carriage are respectively 3 and 4 feet in diameter. Will the two points that are uppermost at the start again be uppermost at the same time, if the carriage should travel on a straight line?

125. Show that in the accompanying figure, d will be a side of an inscribed decagon, and p will be a side of an inscribed pentagon.

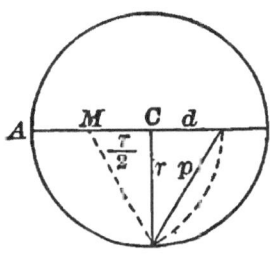

Fig. 225.

THEOREMS.

126. From any point within a regular polygon, if perpendiculars be let fall on the lines forming the sides, their average length will be that of the apothem of the polygon.

127. If on each of the three sides of a right triangle, as diameters, semicircles be constructed, as indicated in the figure, the sum of the areas of the two crescents will equal the area of the right triangle.

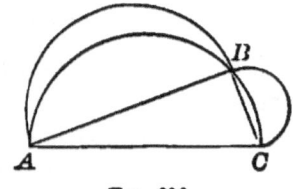

Fig. 226.

128. If four circles are constructed as in the figure, the radius of the small circles will be one-third of the radius of the large circle.

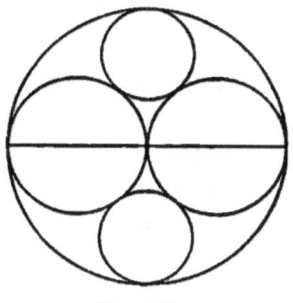

Fig. 227.

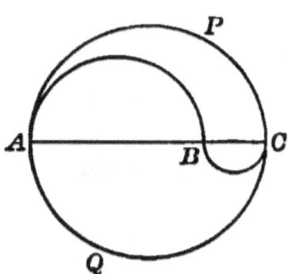

Fig. 228.

129. If the diameter of a circle be separated into any two parts and semicircles be constructed as indicated,

(a) $\widehat{AB} + \widehat{BC} = \widehat{APC}$.

(b) The areas into which the circle is separated by the arcs AB and BC will be to each other as AB is to BC.

130. The area of the ring bounded by two concentric circumferences equals the area of a circle, the diameter of which is a chord of the larger that is tangent to the smaller.

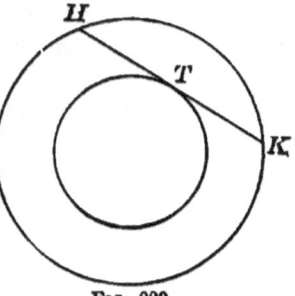

Fig. 229.

131. If $\overline{CA}$ be separated into any number of equal parts, perpendiculars at the points of division be drawn to intersect the circumference constructed on $\overline{CA}$ as a diameter, and these points of intersection be joined with C, they will be radii that will determine rings of equal area.

132. If the centres A and B of two circles be joined, and a circle of any radius a, with the middle of $\overline{AB}$ as its centre, be constructed, the sum of the squares of the tangents to the first circles from any point in the circumference of the third circle, will be constant.

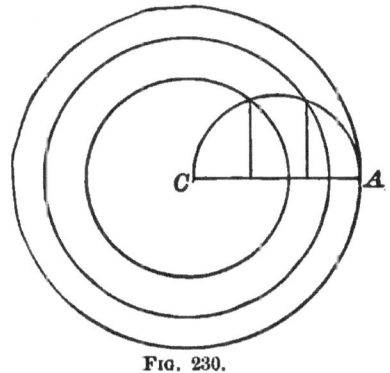

Fig. 230.

133. If G represent the point of intersection of the medians of a $\triangle ABC$, and P be *any* point in the plane,

$$\overline{PA}^2 + \overline{PB}^2 + \overline{PC}^2$$
$$= \overline{AG}^2 + \overline{BG}^2 + \overline{CG}^2$$
$$+ 3\,\overline{PG}^2.$$

NOTE. — G is the centre of gravity of the triangle.

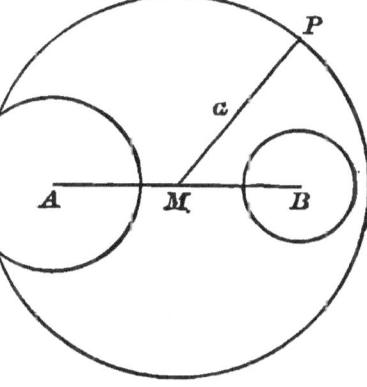

Fig. 231.

134. If any number of parallels intersect two straight lines, the middle points of these parallel segments, together with the points of intersection of the diagonals of the trapezoids formed, all lie in the same straight line.

135. In any triangle, the middle of each side, the feet of the three altitudes, and the middle points of the segments of the altitudes between their common intersection and the vertices of the triangle, are 9 points in the circumference of a circle.

NOTE. — This is one of the famous theorems.

Suggestion. — A simple method of establishing the theorem is by the use of quadrangles.

Show that the circle which passes through

(*a*), 1, 2, 3, will pass through 4.
(*b*), 2, 3, 4, will pass through 5.
(*c*), 2, 3, 5, will pass through 6.
(*d*), 1, 2, 3, will pass through 7.
(*e*), 2, 5, 6, will pass through 8.
(*f*), 2, 3, 4, will pass through 9.

Find the centre.

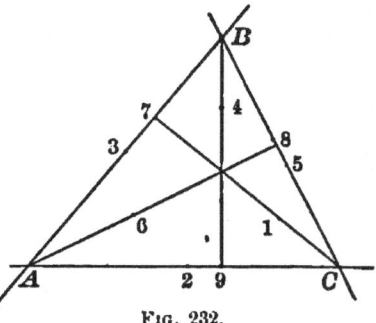

Fig. 232.

PROBLEMS IN LOCI.

136. Find the locus of a point which moves so that the ratio of its distances from two fixed points always remains the same.

Let A and B represent the two fixed points. By Ex. 116, two positions will lie on AB, one between A and B, and the other outside $\overline{AB}$.

Let $\frac{m}{n}$ represent the fixed ratio. If $m > n$, the points on the line will lie as indicated at D and C.

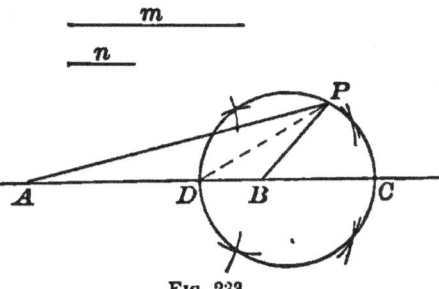

Fig. 233.

$$\frac{DA}{DB} = \frac{m}{n} \text{ and } \frac{CA}{CB} = \frac{m}{n}.$$

A number of constructions made under the given conditions suggest that the locus is the circumference of a circle on $\overline{DC}$ as a diameter.

Analysis. — *If* the suggested figure be the required locus, *any* point on the circumference must fulfil the required conditions and any point not on the suggested locus must fail to fulfil the required conditions; *i.e.* if P be any point on the circumference, we must show that $\overline{PA} \div \overline{PB} = m \div n$; and that if Q be any point not on

PROBLEMS. 171

the circumference, we must show that $\overline{QA} \div \overline{QB}$ does not equal $m \div n$, ($\overline{QA} \div \overline{QB} \neq m \div n$).

Furthermore, if the suggested locus be the locus, any line through D (other than a tangent to the suggested circle) will have D and another point, from which segments of lines drawn to A and to B will have the ratio of $m \div n$.

Demonstration. — Draw any line through D, as DH.

Draw $\overline{BF} \perp$ to DH.

Join A and F.

$$\frac{DA}{DB} = \frac{m}{n}.$$

$$\frac{FA}{FB} > \frac{m}{n}$$

(FA being $> DA$, and FB being $< DB$)

By hypothesis $\frac{m}{n} > 1$.

If a point should move from D toward F, and beyond to infinity, the ratio of its distances from A and B, after passing F, would approach 1, as the limit of the ratio. Then somewhere between F and ∞ the ratio would be $m \div n$.

Let H represent the point.

Draw HA and HB.

The $\angle AHB$ is bisected by HD.

A perpendicular to HD at H will pass through C.

If there be any point Q, other than D and H on the line DH such that

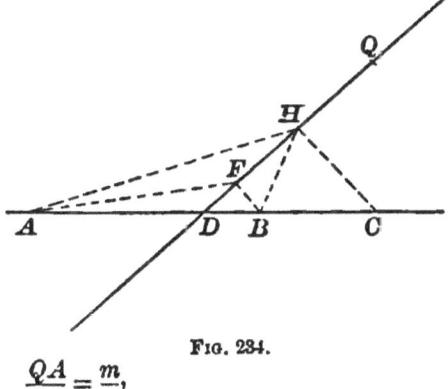

Fig. 234.

$$\frac{QA}{QB} = \frac{m}{n},$$

a perpendicular to HD erected at Q would pass through C, which is impossible.

Hence H will be on the locus of the vertex of a right triangle having DC for its hypotenuse. That locus is the circumference of a circle having DC for its diameter; any point P on the circle in Fig. 1 will be the point H of the line joining P and D; and the suggested circumference is the required locus.

Discussion. — If the ratio $m \div n$ should be made to approach 1, the diameter of the circle, which is the locus, would increase, D would approach the middle of the segment AB, and C would approach ∞. If the ratio should reach 1, D would be at the middle of $\overline{AB}$, C would be at ∞, and what was a circle would reach the limit toward which it approached; viz. a perpendicular bisector of the segment AB. If the ratio fall below 1, the locus would be a circumference having its centre on the opposite side of A.

The centre of the locus depending on the ratio $m \div n$ may occupy any position on the line AB except between the points A and B.

The ratio $m \div n$ may be changed by causing either the numerator or the denominator to change, or both.

If n should remain fixed, and m should increase, both D and C would approach B; and if m should become ∞, the points D and C would coincide with B, and the circle would be reduced to a point. If m should remain fixed, and n should approach 0, the same thing would result.

$$\text{If } \frac{m}{n} < 1,$$

and m should decrease while n remained fixed, the circle which now would encompass A would decrease in diameter, and the circle would degenerate to a point at A when m should become 0. If m should remain fixed and n should approach ∞, the same thing would result.

NOTE. — This is the famous "Problem of the Lights." In this solution the case has been that of the locus of the points of equal illumination on a plane which passes through the line of two luminous points.

In the Solid Geometry it will be shown that the complete locus is the surface of a sphere enveloping the lesser light. The diameter of the sphere is determined by the distance apart of the luminous points and by the ratio $m \div n$.

This problem is very prettily handled by the methods of the Analytic Geometry.

PROBLEMS.

137. Given two circles. Find the locus of a point which moves so that the difference of the squares on the tangents from it to the two circles shall be constant.

NOTE. — When the difference is 0, the locus is called the *radical axis*.

138. Find the locus of a point from any position of which two given circles subtend the same angle.

139. Two circumferences intersect at H and K. Find the locus of the middle point of $\overline{AB}$ through H.

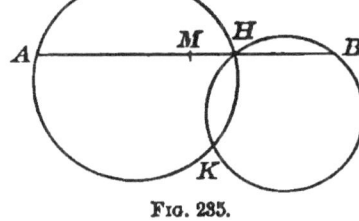

FIG. 235.

140. A circle, the *diameter* of which equals the radius of a larger circle, is internally tangent to the larger one and rolls on its circumference.

Find the locus of any given point P on the circumference of the smaller circle.

Query. — If segments of the rolling circumference be of different colors, what will be the effect of a very rapid rolling of the smaller circle on the larger one?

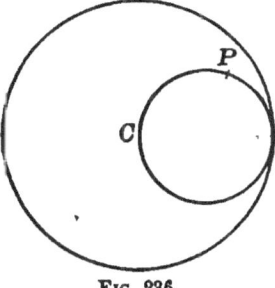

FIG. 236.

NOTE. — This is a special case of a general problem; viz. Find the locus of any point in the plane of a circle, as the latter rolls on the circumference of another circle.

The general case is better treated by the methods of Algebraic Geometry.

141. Find the locus of a point, the distances of which from two given straight lines have a fixed ratio.

142. Find the locus of a point which moves so that the tangents from it to two given circles are equal.

143. Find the locus of a point which moves so that the sum of its distances from two vertices of an equilateral triangle shall equal its distance from the third.

144. Find the locus of a point which moves so that the sum of the squares of the distances from two given points is fixed. The same for three given points.

145. Find the locus of a point which moves so that the difference of the squares of its distances from two fixed points is constant.

146. Within a circle two chords at right angles to each other intersect in a fixed point. Find the locus of the middle points of the sides of the quadrangle of which the perpendicular chords are the diagonals.

SOLID AND SPHERICAL GEOMETRY.

CHAPTER IX.

The first eight chapters of this work have dealt chiefly with the relations of figures in a single plane. The remainder, making use of what has been developed in the Plane Geometry, will not be confined to a plane.

105. THEOREM. *The intersection of two planes is a straight line.*

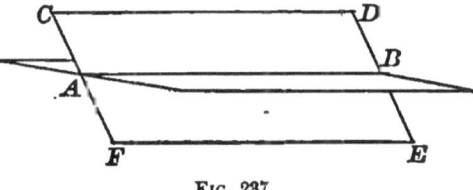

FIG. 237.

A plane has been defined as a surface such that if any two points in it be joined by a straight line, the line will be wholly in the surface.

If two planes intersect, they will have more than one point in common. Any two of these points determine a straight line, which straight line will lie in both planes, hence must be common to both and so be their intersection. Q. E. D.

Planes are usually represented by a quadrangle which is a part of the surface, and the quadrangle is usually taken in the form of a parallelogram, and is designated as the plane *CE*, for instance.

Exercises. — 1. Show that a straight line and a point determine a plane.

2. Show that three points determine a plane.

3. Show that a pair of intersecting straight lines determine a plane.

106. When two planes intersect, they form four diedral angles.

A diedral angle between two planes is the amount of rotation that one plane would have to undergo about the line of intersection as an axis, to coincide with the other.

The measure of a diedral angle is the plane angle formed by two lines, one in each plane, perpendicular to the line of intersection at the same point.

Let *MN* and *QS* represent two planes, *IJ* the line of intersection, *B* any point in that line, *BA* a line in the plane *MN* ⊥ to *IJ*, and *BC* a line in the plane *QS* ⊥ to *IJ*.

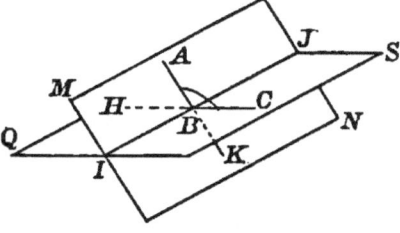

Fig. 238.

The plane ∠ *ABC* is the measure of the diedral *M–IJ–S*.

The vertical plane ∠ *HBK* is the measure of the diedral, vertical to the one measured by *ABC*.

The plane angle *ABH*, the supplement of *ABC*, is the measure of the diedral *M–IJ–Q*.

CBK, the vertical of *ABH*, is the measure of *S–IJ–N*, the vertical of *M–IJ–Q*. Hence the

Theorem. *Vertical diedrals are equal.*

INTERSECTIONS OF PLANES. 177

107. Diedrals are acute, right, or obtuse according as the measuring plane angle is acute, right, or obtuse.

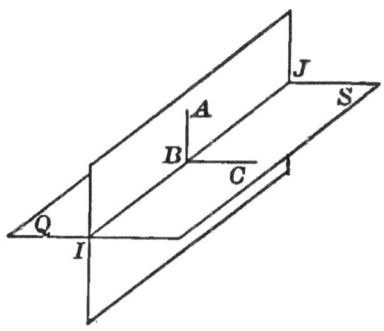

FIG. 239.

If a diedral be 90°, the line BA will be perpendicular to both BI and BC, which are lines on the plane QS.

THEOREM. *If a line be perpendicular to two lines of a plane at their intersection, it will be perpendicular to any line of the plane passing through its foot.*

Let BA be perpendicular to BC and BH. Let BP be any line of the plane QS passing through B.

Analysis. — If BP is perpendicular to AK and only under that condition, we know that oblique lines drawn from any point of BP to points equally distant from the foot of the perpendicular will be equal.

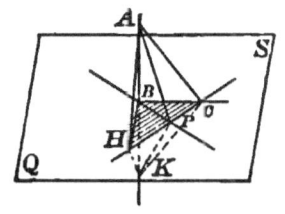

FIG. 240.

Demonstration. — Draw HC, a straight line in the plane QS intersecting the three lines of the plane at H, P, and C. Lay off $BK = BA$ and join the points H, P, and C with A and K.

N

$$HA = HK \quad (\triangle HBA = \triangle HBK),$$
$$CA = CK \quad (\triangle CBA = \triangle CBK),$$
$$HC = HC.$$
$$\therefore \triangle AHC = \triangle KHC,$$
and
$$\angle ACH = \angle KCH.$$
Then
$$\triangle ACP = \triangle KCP$$
(having two sides and included angle in each equal).
$$\therefore PA = PK,$$
and that which was necessary and sufficient to establish the theorem is proved. Q. E. D.

Exercises. — 1. Show that if a line perpendicular to another line which it intersects, be rotated about the second line as an axis, it will generate a plane perpendicular to the axis.

2. Find the locus of a point the ratio of the distances of which from two fixed points equals *one*.

108. Definition. A straight line which is perpendicular to all the lines of a plane passing through its foot is said to be perpendicular to the plane.

We have seen by § 107 that there may be a straight line perpendicular to a plane, passing through a point in the plane. It now remains to show that there can be but *one* such perpendicular through any point.

If PA be a perpendicular, and if another perpendicular could also be drawn through P to the plane MN, as PB, we should have, by taking the plane of the intersecting lines, a plane intersecting the given plane in the line PC. PC being a line in the plane MN is perpendicular to both PA and PB.

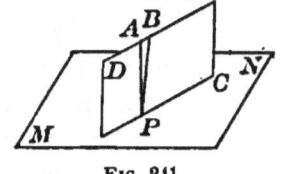

Fig. 241.

But considering simply

the plane *DC*, we know from plane geometry that there cannot be two perpendiculars to a line at the same point. Hence the supposition that a second perpendicular could be drawn to the plane at the point *P* is not allowable, and we have the

THEOREM. *Through any point in a plane* ONE *perpendicular, and* ONLY *one, can be drawn to the plane.*

Exercises. — 1. Prove that any plane containing a perpendicular to a given plane will also be perpendicular to the given plane.

2. Show that if two planes are perpendicular to each other, a line perpendicular to one plane at a point on the line of intersection (*trace*) of the two planes will lie in the other plane.

3. Show how a carpenter's square may be used to erect a perpendicular to a plane at any point in the plane.

4. Show that a triangle, the sides of which are 6, 8, and 10 (or any multiples of 6, 8, and 10) may be used to erect a perpendicular at a point in a plane.

5. Show that through any point not in a plane, *one*, and *only one*, perpendicular can be drawn to the plane.

6. Establish the converse of the theorem in this article.

7. Prove that the perpendicular from a point to a plane is the shortest distance from the point to the plane.

8. Show that equal oblique lines will fall at equal distances from the foot of the perpendicular.

9. Show that the *converse* of Ex. 8 is true.

10. Show that the *opposite* of Ex. 8 is true.

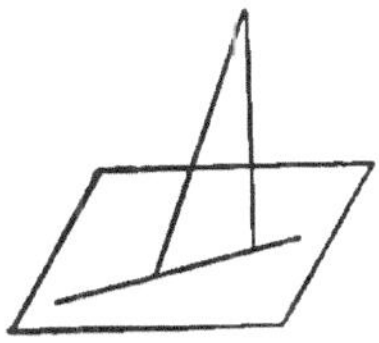

FIG. 242.

11. Show how to construct a perpendicular to a plane from a point without the plane.

12. Show that if two planes make a right diedral, and a perpendicular be drawn to one plane from any point in the other, the foot of the perpendicular will lie in the trace.

180   ELEMENTS OF GEOMETRY.

13. Show that through any line in a plane, *one plane*, and *only one*, can be constructed perpendicular to the given plane.

14. Find the locus of a point when in every position it is equally distant from three given points.

15. Find the locus of a point equally distant from all points of the circumference of a circle.

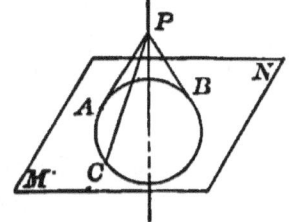

Fig. 243.

109. The orthographic projection of a point on a plane is the foot of the perpendicular to the plane.

There may also be oblique projections. In fact, any straight line through the point, and piercing the plane, will project the point upon the plane at the point of piercing. But when we use the word *projection*, without any modification, we mean the foot of the perpendicular through the point to the plane. This perpendicular is called the **projecting line**.

The projection on a plane, of a line, straight or curved,

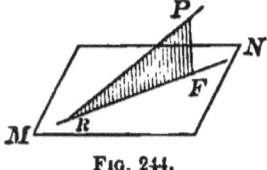

Fig. 244.

Fig. 245.

will usually be a line, straight or curved. If a moving perpendicular to the plane be made to pass through the line that is to be projected, the locus of the foot of the perpendicular will be the projection of the line.

THEOREM. *The projection of a straight line on any plane will be a straight line.*

Let *MN* represent a plane, and *PR* the straight line to

PROJECTIONS OF LINES.

be projected. From any point, as P, let fall the perpendicular PF. The plane determined by the two lines PR and PF will be the perpendicular to the plane MN (Ex. 1, § 108).

This plane (called the **projecting plane**) will contain all points of PR, will contain the projecting line in each of its positions (Ex. 2, § 108), and its trace, which is a straight line, will contain the feet of the perpendiculars.

There will be one projecting plane, and hence but one projection. Q. E. D.

COROLLARY. — The converse is not necessarily true; for if the plane AB make a right diedral with the plane MN, any curve lying in the plane AB will be projected into the trace AC, which is a straight line.

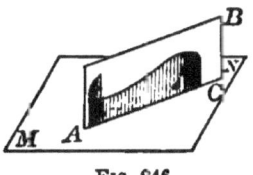

FIG. 246.

Only *plane* curves can be projected in straight lines.

110. THEOREM. *A straight line oblique to a plane makes with its projection both larger and smaller angles than with any other line of the plane through the piercing point.*

Let PR be any oblique line, RF its projection, and RH any line of the plane through R.

Analysis. — If the $\angle PRF$ be less than the $\angle PRH$, and we should lay off RH equal to RF and draw PH, we would have

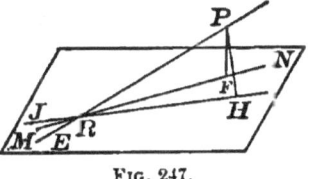

FIG. 247.

two $\triangle$ PRF and PRH having two sides of one equal to two sides of the other and the included angles unequal, and (Ex. 3 at the end of Chapter III.) the third side PH would be greater than the third side PF.

Demonstration. — PH is greater than PF (being an oblique line from P to the plane MN). Hence the ∠ PRF is less than the ∠ PRH.

But every straight line that intersects another straight line makes with it adjacent angles that are supplementary.

Hence ∠ PRM > ∠ PRJ.   Q. E. D.

NOTE. — If the *analysis* were omitted, much printing might be saved, but the solution of the problem would lose the greater part of its educational value.

111. THEOREM. *If two straight lines are perpendicular to the same plane, they will be parallel.*

Let PR and QS represent the perpendiculars.

The analysis suggests the following

Proof. — A plane through PR will be perpendicular to MN; a plane through QS will be perpendicular to MN. Each plane may be so placed that they shall have the common trace PQ. They must then coincide (Ex. 13, § 108).

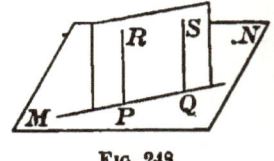

FIG. 249.

The lines PR and QS therefore lie in a plane; and as both are perpendicular to PQ, a line of the plane, they are parallel.   Q. E. D.

Exercises. — 1. Establish the converse if you can.

2. Will the opposite of the theorem be true?

3. Show that if one of two parallels is perpendicular to a plane, the other will be also.

4. Show that if each of two non-parallel planes is perpendicular to a third plane, their line of intersection will be perpendicular to that plane.

112. Theorem. *If from the foot of a perpendicular to a plane, a line be drawn in the plane perpendicular to any line of the plane, and if its intersection with this line be joined with any point in the original perpendicular, the last line drawn will be perpendicular to the line in the plane.*

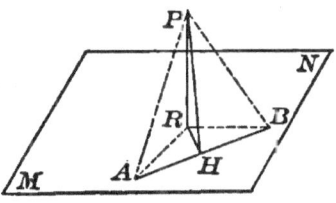

Fig. 249.

Let PR represent the given perpendicular, RH the perpendicular to AB, and PH the line drawn joining H with any point in the perpendicular.

The analysis suggests the

Proof. — On AB, lay off equal distances from H in opposite directions, to A and B. Join A and B with P and with R.

$\triangle RHA = \triangle RHB.\quad \therefore RA = RB.$
$\triangle PRA = \triangle PRB.\quad \therefore PA = PB.$
$\triangle PAH = \triangle PBH.\quad \therefore \angle PHA = \angle PHB.$
$\qquad = 90°.$ Q. E. D.

113. Through any point in a plane, one, and only one, parallel can be drawn to a line of that plane. p. 33.

If a plane be rotated about any line of the plane, and the rotation be continued until the plane returns to its initial position, *every* point in space will have been encountered *once* by *each part* of the rotating plane.

It is thus seen that *one*, and *only one*, parallel to a line can be drawn through any point in space.

184 ELEMENTS OF GEOMETRY.

Theorem. *If a line not in a given plane be parallel to a line of the plane, the line not in the plane will at all points be at the same distance from the plane.*

Let AB represent the line not in the plane, and CD the line in the plane to which AB is parallel. The analysis suggests the following

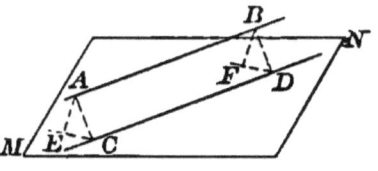

Fig. 250.

Demonstration.—Draw AC and BD, perpendiculars to the two parallels. In the plane MN draw CE and $DF \perp$ to CD at C and D.

The plane determined by AC and CE will be perpendicular to CD and AB. So with the plane determined by BD and DF.

From A and B, perpendiculars to the plane, MN will fall in CE and DF.

$$\triangle ACE = \triangle BDF,$$

since $AC = BD$, $\angle ACE = \angle BDF$, and each are right triangles.

$$\therefore AE = BF. \qquad \text{Q. E. D.}$$

Note.—The line AB is said to be parallel to the plane.

Exercises.—1. Show that if a line EF be drawn, it will be parallel to CD.

2. Show that any plane through AB will intersect the plane MN (if it intersect it at all) in a line parallel to AB.

3. Show that if two lines are parallel to a plane and at the same distance from it, they will lie in the same plane, which will have all its points equally distant from the given plane.

Note.—The two planes in Ex. 3 are called parallel. They are said to never intersect. Also they are sometimes said to intersect in a line at infinity. Both statements mean the same thing.

PROBLEMS.

114. Problems. — 1. Show that if two parallel planes are intersected by a third plane, the lines of intersection will be parallel.

2. Show that if one of any number of parallel planes be cut by a plane, all will be, and the lines of intersection will be parallel.

3. Show that two or more lines which pierce three parallel planes will be divided proportionally.

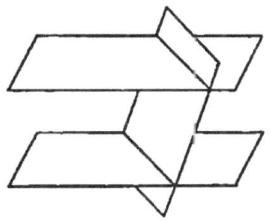

Fig. 251.

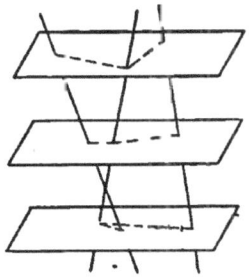

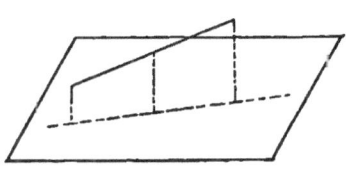

Fig. 252.

4. Show that the projecting line of the middle point of a segment that is projected on a plane will be the half-sum of the projecting lines of the segment extremities.

5. If from any point within a diedral, perpendiculars be let fall on the planes forming the diedral, the angle formed by the perpendiculars will be the supplement of the diedral.*

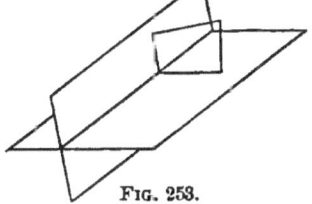

Fig. 253.

6. Show that a plane may be passed through a line parallel to a given line; and that a plane may be passed through a point parallel to two given lines.

---

* When two planes intersect, they make angles that are supplementary to each other, just as two intersecting lines do.

7. Two lines in space which are not parallel may be joined by *one*, and *only one*, mutual perpendicular; and the perpendicular will be shorter than any other distance between points on the two lines.

Let AB and CD represent the two lines.

(a) *Analysis.* — If the lines have a mutual perpendicular, and PR represent it, it would be perpendicular to AB and to RK — a line through $R \parallel$ to CD — and therefore perpendicular to the plane of AB and RK. (Let MN represent this plane, and RK the projection of CD on the plane.)

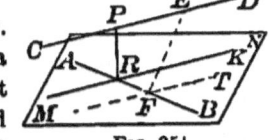

Fig. 254.

Demonstration. — Through one of the lines pass a plane parallel to the second line. Project the second line upon this plane, and at its point of intersection with AB erect a perpendicular to the plane. Being perpendicular to the plane, it will be perpendicular to AB and RK. And being perpendicular to RK, it will be perpendicular to CD. Hence a perpendicular can be drawn joining two lines in space.

(b) *Analysis.* — If a second mutual perpendicular could be drawn, and FE represent it, a line FT, through $F \parallel$ to CD, would also be a perpendicular to FE, and the plane of the lines AB and FT will be perpendicular to FE. But the plane of the lines AB and FT would be the same as the plane of the lines AB and RK. Hence FE would be perpendicular to MN, and F would be the projection of the point E upon the plane MN.

Demonstration. — Through one of the lines pass a plane parallel to the second line. Project the second line upon the plane.

Our analysis shows that *if* FE be a mutual perpendicular to AB and CD, the line FT must be the projection of CD. But by § 109, a straight line can have but one projection on a given plane. Therefore FE cannot be a mutual perpendicular, and there can be but one mutual perpendicular to two lines in space.

Discuss the problem — including the cases when the two lines are parallel and when they intersect.

8. Show that if a line be perpendicular to one of two intersecting planes, its projection upon the other plane will be perpendicular to the line of intersection of the two planes.

9. Show that if two angles in space have their sides parallel and extending in the same direction, the angles will be equal and the planes of the angles will be parallel.

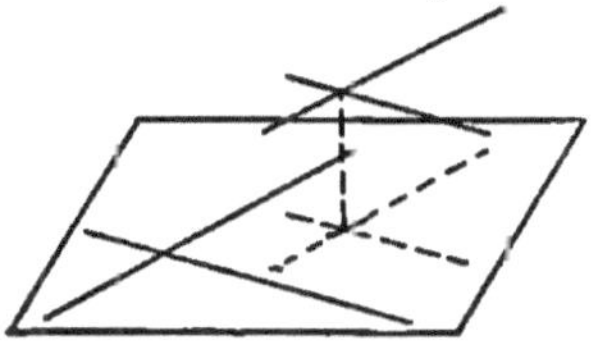

10. Show that the locus of a point, the ratio of the distances of which from two given planes, is ± 1, will be the plane bisectors of the diedral angles, and will themselves form a right diedral.

Fig. 255.

CHAPTER X.

115. Definitions. 1. The figure formed by a surface, all points of which are equally distant from a fixed point, is called a **sphere**.

2. The fixed point is called the **centre**.

3. The surface completely encloses a portion of space.

4. The surface is the locus of a point at a fixed distance from a given point.

Theorem. *Every plane section of a sphere is a circle.*

Let C represent the centre of the sphere, and P, Q, and R points in the section.

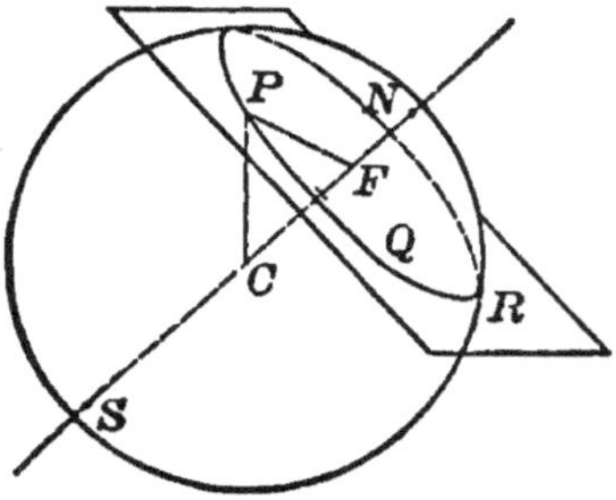

Analysis. — *If* the plane section *be* a circle, it will have a centre, and if at that centre a perpendicular to the plane be erected, it will pass through the centre of the sphere (see Ex. 15, § 108).

Fig. 256.

Demonstration. — From C let fall a perpendicular CF upon the plane. Join P with F and C. Rotate the right $\triangle CPF$ about CF as an axis. The point P as it moves will be at a fixed distance from C, and so must be on the surface of the sphere; it will also be in the secant plane because FP will be. The point P will therefore move over the line of intersection of the two surfaces.

But the locus of a point in a plane at a given distance from a fixed point is the circumference of a circle. Q. E. D.

Definitions. 1. The points N and S, where the line CF pierces the surface of the sphere, are called **poles** of the circle PQR, and the line NS is called its **axis**.

2. If the secant plane pass through the centre of the sphere, the circle of intersection is called a **great circle**—otherwise, a **small circle**.

Exercises. — 1. Show that all great circles of a sphere are equal, and that the intersection of any two will be a diameter of the sphere.

2. Show that a plane section not through the centre of a sphere will determine a circle of smaller radius than a great circle.

3. Show that the greater the distance from the centre that the secant plane is passed, the smaller will be the circle of intersection.

4. Show that if a circle be revolved about a diameter as an axis, it will generate a sphere.

5. Show that the pole N or S is equally distant from all points of the circumference of the circle of intersection.

6. Show that *in general, one*, and *only one*, great arc may be passed through two points on the surface of a sphere.

7. Show that, in general, an infinite number of arcs of small circles may be passed through two points on the surface of a sphere.

8. Making use of the facts determined in Ex. 5, § 103, show that the arc of a great circle joining two points on the surface of a sphere will be shorter than the arc of a small circle joining them.

Remark.—This fact plays an important part in ocean navigation.

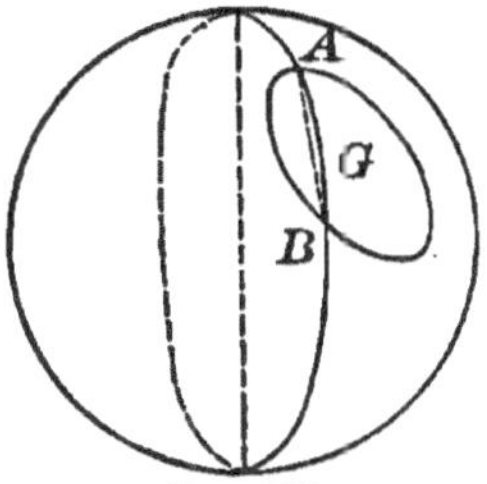

Fig. 257.

9. An arc of a great circle which subtends an angle of 90°

is called a quadrant. Show that all points of a great circle are at a quadrant's distance from its poles.

10. A plane through its centre bisects the surface of a sphere and also its volume.

116. Definition. The angle between any curves which intersect is the angle made by the tangents to the curves at the point of intersection.

Let two great circles intersect on the diameter AX. Through the centre pass a plane perpendicular to AX; and at A draw tangents to the great circles.

$\angle HCK = \angle H\text{-}AX\text{-}K,$

$\angle NAM = \angle KCH = \widehat{KH}.$

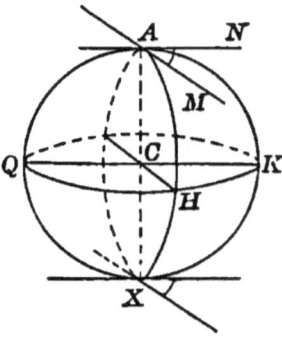

Fig. 258.

The angle between the arcs HA and KA is said to be measured by the arc KH, of a great circle included between its sides and 90° from the vertex.

Exercises.—1. Show that the poles of all great circles which pass through AX will be found in the circumference KHQ, the plane of which is perpendicular to AX at its middle point.

2. Show how, at a given point of an arc, to erect an arc perpendicular to the given arc.

3. Through any two points show how to pass the arc of a great circle.

4. Show how, through a point on a sphere, to let fall a perpendicular to a given great circle.

Definitions.—1. Circles having the same poles are said to be parallel.

2. The portion of the surface of a sphere included between parallel planes is called a *zone*.

3. The volume corresponding to a zone is called a *spherical segment.*

SPHERICAL ARCS.

NOTE. — A spherical blackboard is a most useful adjunct to the study of spherical geometry. A diameter of 20 inches is a convenient size. A quadrant made of wood is also a great convenience. If a regular spherical blackboard of the size suggested cannot conveniently be had, a papier-mâché globe can be slated over and made to serve the purpose fairly well.

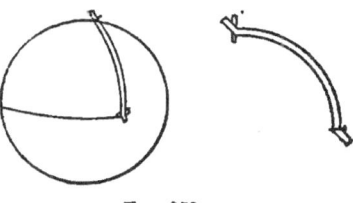

FIG. 259.

5. Show that if a great circle and a small circle have the same poles, they will intercept equal arcs of the great circles which pass through the poles.

6. Show that two small circles having the same poles will intercept equal arcs of the great circles which pass through the poles.

7. Show that if two great circles intersect as in AX, and from points of one of them perpendiculars be let fall to the other, they will vary in length from 0 to the measure of the angle, as the point from which the perpendiculars are let fall passes from A to H.

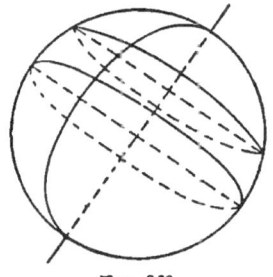

FIG. 260.

8. If AB is tangent to a given great circle ETD at T, and if through AB any plane be passed, it will, in general, intersect the

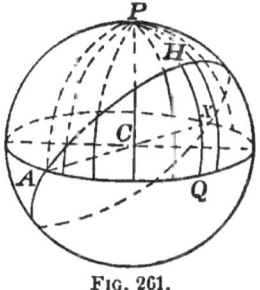

FIG. 261.

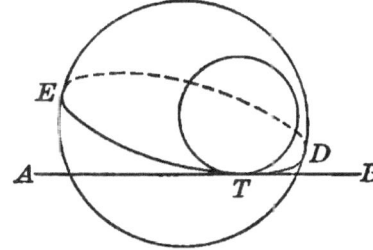

FIG. 262.

sphere in a small circle, which will be tangent to the great circle at

T, will lie entirely in one of the hemispheres determined by the great circle, and will not touch the circumference of the great circle in any other point.

Demonstration.—The only portion of the planes of the two circles which are in common, is the line *AB*, of which the point *T* is the only point on the surface of the sphere. *AB* being in the plane of each circle, and touching each at but one point, is tangent to each at that point. Hence the circles are tangent to each other at the same point and it is the only point common to both.

117. Definitions. The two equal parts (Ex. 10, § 115) into which the circumference of a great circle separates the surface of a sphere, are called **hemispheres**.

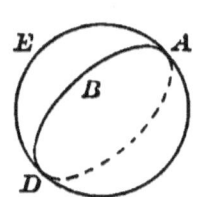

Fig. 263.

The circumferences of two great circles separate the surface of a sphere into four parts, each of which is called a **lune**. The surface included by the semicircles *AED* and *ABD* is a lune.

The circumferences of three great circles, when they do not have a common diameter, separate the surface into eight parts, each of which is called a **spherical triangle** (sph. △).

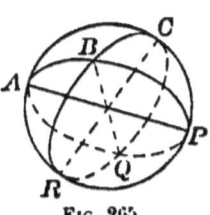

Fig. 264.

If through the vertices (*A*, *B*, and *C*) of a spherical triangle, diameters be drawn, their other extremities (*P*, *Q*, and *R*) determine a spherical triangle which is called the **symmetrical** of the first spherical triangle.

Fig. 265.

Theorem.—*The angles and sides of a spherical triangle and its symmetrical are mutually equal.*

Let *ABD* represent a spherical triangle, and *E*, *F*, *G*, the extremities of the diameters of the sphere through A, B, and D. The sph. $\triangle EFG$ is the symmetrical of the sph. $\triangle ABD$.

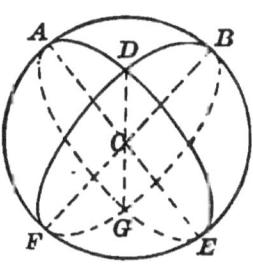

Fig. 266.

$$\angle BAD = \angle GAF$$

(being vertical);

$$\angle GAF = \angle GEF$$

(being the angle between the planes of AFE and AGE).

$$\therefore \angle BAD = \angle GEF.$$

In the same way we may show that

$$\angle ADB = \angle FGE$$

and $\quad\angle ABD = \angle EFG.$

$\overset{\frown}{AB} = \overset{\frown}{EF}$ (each being the supplement of $\overset{\frown}{AF}$),

$\overset{\frown}{AD} = \overset{\frown}{EG}$ (each being the supplement of $\overset{\frown}{DE}$),

and $\overset{\frown}{BD} = \overset{\frown}{FG}$ (each being the supplement of $\overset{\frown}{DF}$).

Q. E. D.

Exercise.—1. Show that if a spherical triangle is isosceles, it may be brought to coincide with its symmetrical.

NOTE.—In general the spherical triangles that are symmetrical cannot be superimposed, because the equal parts are not arranged in the same order, and if one figure be reversed, the curvature will not permit of coincidence of surface.

The plane triangle that has its parts equal, but differently arranged, from another triangle, may be reversed and the two brought to coincide.

The fact that symmetrical spherical triangles are equivalent in area will be established later. At present we are only concerned with the facts presented in the theorem.

o

118. Theorem. — *If at the middle point of a given segment of a great circumference, a perpendicular great arc be erected, all points in the perpendicular arc will be equally distant from the extremities of the segment.*

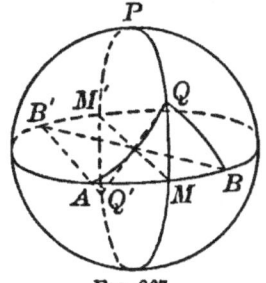

Fig. 267.

Let AB represent the arc segment, of which M is the middle point. Let MQP represent the perpendicular arc, and Q any point of it.

Analysis. — If $QB = QA$, the sph. $\triangle\, QMB$ and QMA would have the three sides of one equal to the three sides of the other, and the angles at M in each equal, but we are unable to substitute the one for the other and show that they are identical. This suggests the use of the symmetrical spherical triangle. Hence the

Demonstration. — The diameters through Q, M, and B determine the symmetrical sph. $\triangle\, Q'M'B'$. Then if the symmetrical sph. $\triangle\, Q'M'B'$ be caused to slide over the surface of the sphere, rotating 180° about the axis of the $\odot MQP$,

$\widehat{M'Q'}$ will coincide with $\widehat{MQ}$,

the rt. $\angle Q'M'B'$ " " " $\angle QMA$,

$\widehat{M'B'}$ " " " $\widehat{MA}$,

and $\widehat{Q'B'}$ " " " $\widehat{QA}$

(because between two points on the surface of a sphere one, and only one, great arc can be drawn).

But $\widehat{Q'B'} = \widehat{QB}.$

$\therefore \widehat{QA} = \widehat{QB}.$ Q. E. D,

SPHERICAL ARCS. 195

119. Theorem. — *From any point on the surface of a sphere, the perpendicular arc (less than 90°) to any great circle will be less than any oblique arc.*

The analysis suggests that if with Q as a pole and $\widehat{QM}$ as an arc radius, a small circle be constructed on the surface of the sphere, it will be tangent to the great circle at M and will have no other point in common with it (Ex. 8, § 116);* and as the arc-distance from Q of every point in the circumference of the small circle is the same as QM,

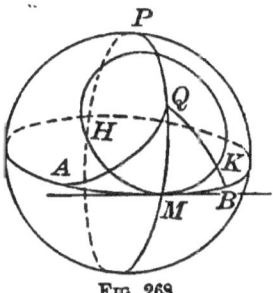

Fig. 269.

$$\widehat{QM} = \widehat{QK} < \widehat{QB}.$$
Q. E. D.

Exercises. — 1. Show that QPH will be greater than any oblique arc from Q to the great circle ABH.

2. Show that of two oblique arcs from Q to the great circle ABH, the one that falls at the greater distance from M will be the greater.

3. Show that as the oblique arcs increase, the angle made by them with the base of the hemisphere will *decrease* from 90° until the oblique arc shall have reached 90° (when the angle will be measured by the arc MQ), and that thereafter the angle will increase until the perpendicular arc, QPH, is reached.

---

*Note. — Any chord of a sphere, except a diameter, may be the common chord of a great circle and any number of small circles or of two or more small circles. A small circle may be tangent to another small circle or to a great circle. But a great circle cannot be tangent to another great circle.

The student may be aided in his conception of these relations, by passing a plane through the diameter or chord under consideration and rotating it, observing the spherical sections.

120. Theorem. — *The sum of two sides of a spherical triangle is greater than the third side.*

Let ABC represent the spherical triangle, AC being the largest side.

Analysis. — If $\overset{\frown}{AC}$ be less than $\overset{\frown}{AB} + \overset{\frown}{BC}$, it may be separated into two parts, one of which shall be less than $\overset{\frown}{AB}$ and the other less than $\overset{\frown}{CB}$. Can this be done?

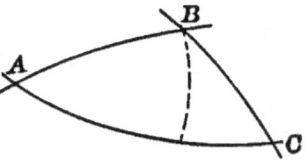

Fig. 269.

The student will furnish the demonstration.

Exercises. — 1. Show that the difference between two sides of a spherical triangle will be less than the third side.

2. It has been shown in § 118 that all points in the perpendicular arc bisecting a given segment will be equally distant from the extremities. Let the student now show that any point not on such perpendicular will not be equally distant from the segment extremities, but will be nearer the extremity on its own side of the perpendicular.

Note. — The student will observe the similarity of method and relation in Plane and Spherical Geometry.

121. Problems. — 1. To construct a small circle the circumference of which shall pass through the vertices of a spherical triangle.

2. Making use of Ex. 1, § 117 and Prob. 1 above, show that two symmetrical spherical triangles are equivalent in area.

3. Show that if the three sides of a spherical triangle be given, the spherical triangle may be constructed.

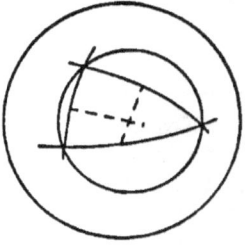

Fig. 270.

It follows that if two spherical triangles have the three sides of one equal to the three sides of the other, the spherical triangles will be superimposable or symmetrical, and so will have the angles as well as sides equal, and will be equivalent in area.

SPHERICAL TRIANGLES. 197

4. To construct on the surface of a sphere an angle equal to a given spherical angle.

Let $\angle ABD$ represent the given spherical angle, and P the point on a given arc QP, where the vertex of the required angle is to be drawn.

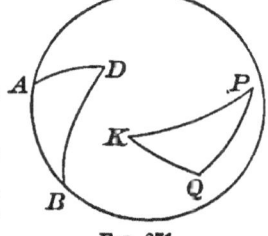

Fig. 271.

The analysis suggests that if any arc AD be drawn, a distance PQ, equal to BA, be laid off from P, either with dividers or tape, and on that as a base, a spherical triangle with sides equal to those of sph. $\triangle ADB$ be constructed (after the manner indicated in Prob. 3), the $\angle QPK$ will equal the $\angle ABD$.

Remark. — It is customary — and convenient — to lay off BA and BD equal; but the principle is of course the same.

5. Show that a spherical triangle will be less than a hemisphere in area.

NOTE. — No side or angle of a spherical triangle may be greater than 180°. If a three-sided figure should be constructed on a sphere, one side of which is greater than 180°, the remainder of the hemisphere is regarded as the spherical triangle.

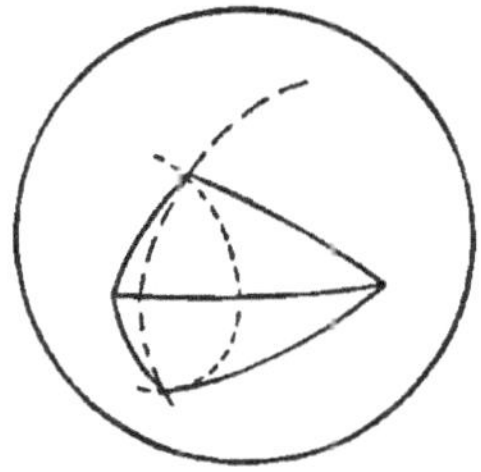

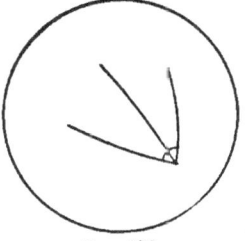

Fig. 272.

If a three-sided figure, one angle of which is greater than 180°, be constructed on a sphere, the arcs will all lie on a hemisphere and the excluded part of the hemisphere is regarded as the spherical triangle, no angle of which will be 180°.

6. Show how to construct a spherical triangle, having given two sides and the included angle.

NOTE. — It follows that if two spherical triangles have two sides and the included angle of one equal to two sides and the included angle of the other, they will be superimposable, or symmetrical, and hence have the other side and remaining angles equal, and areas equivalent.

Fig. 273.

7. Show how to construct a spherical triangle, having given two angles and the included side.

Note. — It follows that two spherical triangles having two angles and included side of one equal to two angles and included side of the other, will have their remaining parts equal and will be equivalent in area.

8. To construct a spherical triangle, having given two sides and an angle opposite one of them.

At any point (A) of a great arc, construct the given angle. On either side of the angle, lay off the given adjacent side ($\stackrel{\frown}{AB}$). With an arc-radius equal to the side that is to be opposite the given angle,

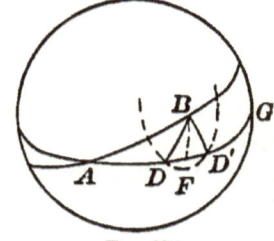

Fig. 274.

and with B as a pole, construct a small circle intersecting $\stackrel{\frown}{AG}$ at D and D'. Joining them with the point B, we will have the *two* sph. △ ADB and $AD'B$, that contain the given parts.

Discussion. — If from B a ⊥ $\stackrel{\frown}{BF}$ be let fall on the $\stackrel{\frown}{AG}$ (Ex. 4, § 116), it will be the shortest distance from B to $\stackrel{\frown}{AG}$ (§ 119).

(a, 1.) If the side opposite the given angle be less than the perpendicular, there will be *no* construction.

(a, 2.) If equal to the perpendicular, there will be *one* construction — the spherical triangle being right-angled.

(a, 3.) If intermediate in value between the perpendicular and $\stackrel{\frown}{BA}$, and also intermediate in value between the perpendicular and $\stackrel{\frown}{BA'}$ (the supplement of $\stackrel{\frown}{BA}$), there will be *two* constructions.

(a, 4.) If intermediate in value between either of the arcs, AB or $A'B$, and the perpendicular, but not between the other one and the perpendicular, there will be but *one* construction.

(a, 5.) If not intermediate in value between the perpendicular and either $\stackrel{\frown}{AB}$ or $\stackrel{\frown}{A'B}$, there will be *no* construction.

What we have seen in the discussion so far has been developed under the supposition that the ∠ BAG was an acute angle. Let us consider the case where it is obtuse.

SPHERICAL TRIANGLES.

The perpendicular from B will pass through P (the pole of $\widehat{AG}$), will be greater than 90°, and will be greater than any other arc drawn from B to any point in $\widehat{AG}$.

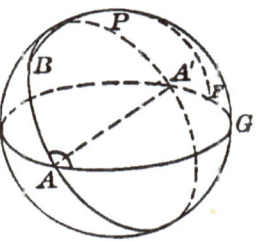

Fig. 275.

(*b*, 1.) If the side opposite the given angle be *greater* than this perpendicular, there will be *no* construction.

(*b*, 2.) If the side be equal to the perpendicular, there will be *one* construction, and the spherical triangle will be right-angled.

(*b*, 3.) If the side be intermediate in value between the perpendicular and $\widehat{BA}$, and also intermediate between the perpendicular and $\widehat{BA'}$, there will be *two* constructions.

(*b*, 4.) If the side be intermediate in value between the perpendicular and but *one* of the arcs AB or $A'B$, there will be *one* construction only.

(*b*, 5.) If not intermediate in value between the perpendicular and either the other given side or its supplement, there will be *no* solution.

NOTE. — Case (*a*, 1) might be included with (*a*, 5), and (*b*, 1) with (*b*, 5); but for the sake of gradual approach it is deemed best to present them in this order.

122. Definitions. If with the vertices of a spherical triangle as poles and quadrants as arc-radii, a spherical triangle be constructed, it is called the polar of the given spherical triangle.

Let ABC represent the given spherical triangle, and DK, KG, and GD, the arcs constructed with B, A, and C as poles.

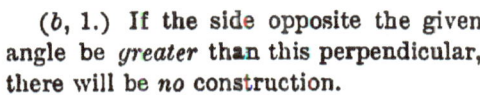

Fig. 276.

Since all points of $\widehat{DK}$ are at a quadrant's distance from B, and all points of $\widehat{GD}$ are at the same distance from C, D will be the pole of $\widehat{BC}$.

For the same reasons, K will be the pole of $\widehat{AB}$, and G will be the pole of $\widehat{AC}$.

Thus it is seen that the relation is a mutual one, and if sph. $\triangle DKG$ is the polar of ABC, the latter is also the polar of the former.

$$\widehat{DM} = 90°,$$
$$\widehat{NK} = 90°.$$
$$\therefore \widehat{DM} + \widehat{NK} = 180°,$$
or $\quad \widehat{DN} + 2\,\widehat{NM} + \widehat{MK} = 180°,$
or $\quad \widehat{DK} + \widehat{NM} = 180°.$
But $\quad \widehat{NM} = \angle B.$ (§ 116)
$$\therefore \widehat{DK} + \angle B = 180°.$$

In the same way it may be shown that:
$$\widehat{KG} + \angle A = 180°,$$
and $\quad \widehat{GD} + \angle C = 180°.$

Again, $\quad \widehat{EC} = 90°,$
$$\widehat{AJ} = 90°.$$
$$\therefore \widehat{EC} + \widehat{AJ} = 180° = \widehat{EA} + 2\,\widehat{AC} + CJ,$$
or $\quad \widehat{EJ} + \widehat{AC} = 180°.$
But $\quad \widehat{EJ} = \angle G.$
$$\therefore \angle G + \widehat{AC} = 180°.$$

In the same way it may be shown that $\widehat{BC} + \angle D = 180°$, and $\widehat{BA} + \angle K = 180°$. Hence the

SPHERICAL TRIANGLES.

THEOREM. *If two spherical triangles are polar to each other, any side of either will be the supplement of the angle in the other at the pole of that side.*

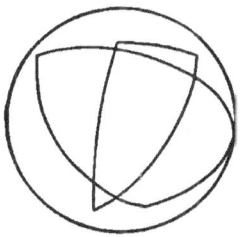

FIG. 277.

NOTE. — If any of the sides of a spherical triangle, or its polar, are greater than 90°, some of the sides of one will cross sides of the other.

123. Problems. — 1. To construct a spherical triangle, having given the three angles.

Analysis. — If we had the required spherical triangle and should construct its polar, the sides of the latter would be the supplements of the three given angles.

Construction. — With the supplement* of each of the given angles as sides, construct a spherical triangle (Ex. 3, § 121). This will be the polar of the required spherical triangle. Construct its polar; it will be the required spherical triangle.

2. To construct a spherical triangle, having given two angles and the side opposite one of them, making use of the polar spherical triangle.

3. To construct a spherical triangle, having given two angles and the side opposite one of them, without using the polar. Make a careful discussion.

4. Show that 0° and 360° are the limits of the sum of the three sides of a spherical triangle.

---

* To get an arc the supplement of an angle, take that part of a semicircumference having the vertex of the angle as its pole, not included by the arcs forming the angle.

5. Show that 180° and 540° are the limits of the sum of the three angles of a triangle.

6. The student will observe that when the three angles are given of a plane triangle, the triangle is undetermined (Ex. 6, § 28), while the three angles of a spherical triangle completely determine it. Show from this that if two spherical triangles have the three angles of each equal, each to each, the spherical triangles are equal or symmetrical.

124. Theorem 1. *If two sides of a spherical triangle are equal, the angles opposite them will be equal.*

Let BA and BC represent the equal sides.

From the middle point M of the arc AC draw the arc MB.

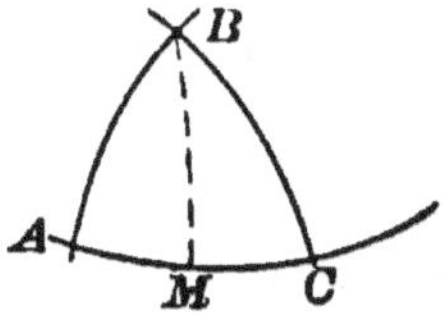

$\triangle AMB = \triangle CMB$ (3 sides equal).

$\therefore \angle A = \angle C$, Q. E. D.

$\angle AMB = \angle CMB = 90°$,

$\angle ABM = \angle CBM$.

Fig. 278.

Theorem 2. *Conversely, if two angles of a spherical triangle are equal, the sides opposite them will be equal.*

If the sides are *not* equal, and $AB > BC$, a perpendicular erected at the middle point of the side included by the equal angles will not pass through the vertex. (The perpendicular is the locus of all points equally distant from A and C.) It will therefore intersect the sides in succession as at H and K. If the auxiliary $\overset{\frown}{HC}$ be drawn, it will make with MC an angle equal to the angle at A, but

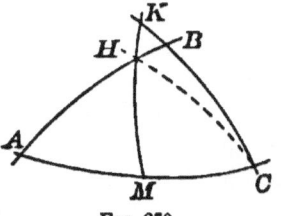

Fig. 279.

smaller than the ∠ BCA. But that is contrary to the hypothesis.

Hence the two sides cannot be unequal; or in other words, they are equal. Q. E. D.

Exercises. — 1. Construct a right spherical triangle, having given the sides adjacent to the right angle.

2. The same, having given the hypothenuse and one side.

3. The same, having given the hypothenuse and an oblique angle.

4. The same, having given an oblique angle and the adjacent base.

5. The same, having given an oblique angle and the side opposite. Discuss thoroughly.

6. Show that two of the angles of a spherical triangle may each be 90°; and that when they are such, the sides opposite them will be 90°.

7. Show that the three angles of a spherical triangle may each be 90°; that when they are such, each side of the spherical triangle will be 90°; and that the area of the spherical triangle will be one-eighth of the surface of the sphere.

8. Show that if two spherical triangles can be brought to have a common side, and two other sides cross each other, the sum of the sides that cross is greater than the sum of the sides that do not cross; i.e.

$$\widehat{AC} + \widehat{BE} > \widehat{AB} + \widehat{CE}.$$

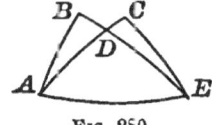

Fig. 280.

9. Show that if two spherical triangles have a common side $\widehat{AC}$, and the vertex D, of one, lies within the other,

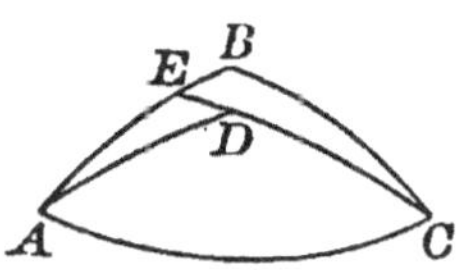

$$\widehat{AB} + \widehat{BC} > \widehat{AD} + \widehat{DC}.$$

10. Show that if two spherical triangles have two sides in one equal to two sides in the other, and the included angles are unequal, the third sides will be unequal, and the greater side will be opposite the greater angle.

Fig. 281.

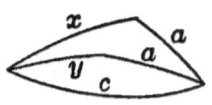 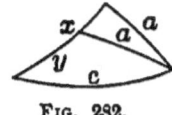 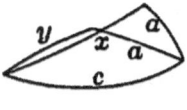

Fig. 282.

11. Show that the sum of the sides of a convex* spherical polygon of any number of sides is less than a great circle.

Let $ABCDE$ represent any spherical polygon. Continue any side, as AB, so as to form a great circle; it separates the sphere into hemispheres. The angles at A and B are less than 180° each. If any of the vertices of the polygon should fall outside of the hemisphere in which $\widehat{AE}$ and $\widehat{BC}$ lie, we should have some of the angles greater than 180°. And in case the

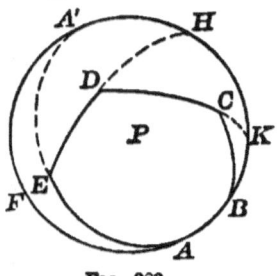

Fig. 283.

polygon were separated into spherical triangles by means of diagonals, some of the spherical triangles would be exterior to the polygon, and the polygon would not be convex. Therefore the spherical polygon lies entirely within the hemisphere.

Produce $\widehat{AE}$ to A', $\widehat{ED}$ to H, and $\widehat{DC}$ to K.

$$\widehat{AFA'} = \widehat{AEA'} = \widehat{AE} + \widehat{EA'},$$
$$\widehat{EA'} + \widehat{A'H} > \widehat{ED} + \widehat{DH},$$
$$\widehat{DH} + \widehat{HK} > \widehat{DC} + \widehat{CK},$$
$$\widehat{CK} + \widehat{KB} > \widehat{CB},$$
$$\widehat{AB} = \widehat{BA},$$

$\widehat{AFA'} + \widehat{EA'} + \widehat{A'H} + \widehat{DH} + \widehat{HK} + \widehat{CK} + \widehat{KB} + \widehat{AB} > \widehat{AE} + \widehat{EA'}$
$\qquad + \widehat{ED} + \widehat{DH} + \widehat{DC} + \widehat{CK} + \widehat{CB} + \widehat{BA};$

$\widehat{AFA'} + \widehat{A'H} + \widehat{HK} + \widehat{KB} + \widehat{AB} > \widehat{AE} + \widehat{ED} + \widehat{DC} + \widehat{CB} + \widehat{BA};$

$\qquad$ Great circle $> \widehat{AE} + \widehat{ED} + \widehat{DC} + \widehat{CB} + \widehat{BA}.$

Q. E. D.

---

*Note.— A convex spherical polygon is one in which none of the angles are greater than 180°.

CHAPTER XI.

125. Definitions. We have seen that a single plane separates space into two parts; two planes separate it into four parts, and themselves intersect in a straight line.

In general, three planes separate space into eight parts, called **triedrals**; have three lines of intersection; and have *one*, and *only one*, common point.

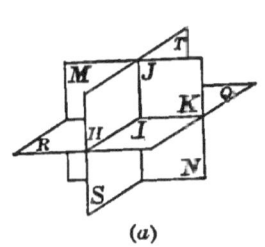 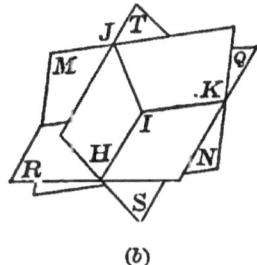

(a)   FIG. 284.   (b)

Fig. (a) represents the three planes so placed as to be perpendicular to each other. In Fig. (b) they are oblique.

Particular cases may arise; for instance:

1. The third plane may be parallel to the line of intersection of the first and second. The three lines of intersection will then be parallel.

In this case space is separated into seven parts.

2. The third plane may pass through the line of intersection of the first and second. In this case space is separated into six parts.

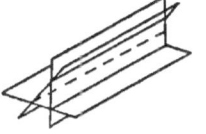

FIG. 285.

3. The three planes may be parallel to each other. In this case they are sometimes said to intersect at infinity. They separate space into four parts.

When three planes form eight triedrals, the common point (*I* in the figure) is called the vertex of each.

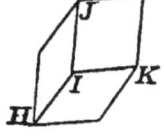

Fig. 286.

The lines of intersection of the planes form the *edges* of the triedrals.

IJ, *IK*, and *IH* are edges.

The angles that the edges make with each other are called the **facial angles**.

The angles that the planes make with each other (as previously noted) are called **diedrals**.

126. Theorem. *If the vertex of a triedral be taken as the centre of a sphere of any radius, the intersections will form a spherical triangle, the sides of which will be the measures of the facial angles; and the angles of which will be the measures of the diedrals.*

With *I* as a centre and *ID* as a radius describe the sphere.

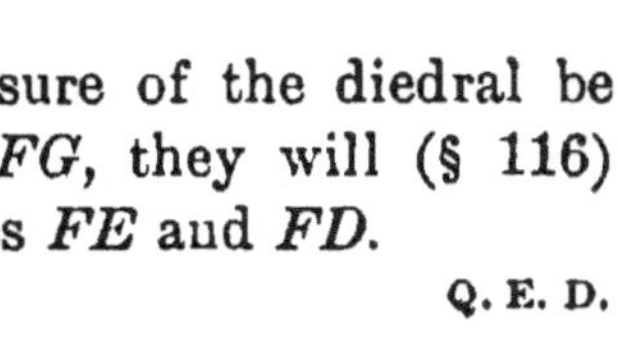

The $\widehat{DE}$ is the measure of the $\angle DIE$; the $\widehat{EF}$ is the measure of the $\angle EIF$, and the $\widehat{FD}$ is the measure of the $\angle FID$.

Fig. 287.

If at any of the vertices of the spherical triangle, as *F*, the measure of the diedral be drawn (§ 106) and be *FO* and *FG*, they will (§ 116) make the same angle as do the arcs *FE* and *FD*.

Q. E. D.

Exercises. —1. Show that if two facial angles of a triedral are equal, the diedrals opposite them will be equal.

Let IJ, IH, and IK represent the edges of the triedral * of which $\angle EID = \angle EIF$.

Conceive a sphere with I as a centre and any radius, say IE, intersecting the planes of the triedral in $\widehat{ED}$, $\widehat{DF}$, and $\widehat{EF}$.

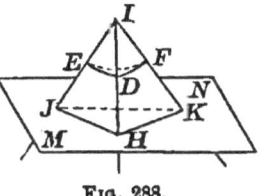

Fig. 283.

Since $\angle EID = \angle EIF$, $\widehat{ED} = \widehat{EF}$. ∴ (§ 124) $\angle EDF = \angle EFD$.

Hence from the theorem, the diedrals J-HI-K and J-KI-H are equal.

2. Show that if two diedrals of a triedral are equal, the facial angles opposite them will be equal.

3. Show that in any triedral the sum of any two facial angles will be greater than the third.

4. Show that the facial angles and the diedrals of a triedral and its vertical triedral will be equal.

5. Show that the sum of the facial angles of a triedral will be between 0° and 360°.

6. Show that the sum of the diedrals of a triedral will be between two and six right angles.

7. Show that two triedrals having the three facial angles of one equal to the three facial angles of the other, and similarly arranged, are superimposable.

127. Definition. If two triedrals can be so placed that the edges of one are perpendicular to the faces of the other, they are said to be *supplementary*.

Let the two triedrals be so placed that their vertices coincide; VK being perpendicular to the face AVC, VE perpendicular to the face AVB, and VD perpendicular to the face BVC.

---

* NOTE.— The plane MN is introduced simply for the purpose of enabling the beginner to more readily comprehend the figure.

A sphere with V as a centre would be intersected by the faces so as to form on its surface two spherical triangles.

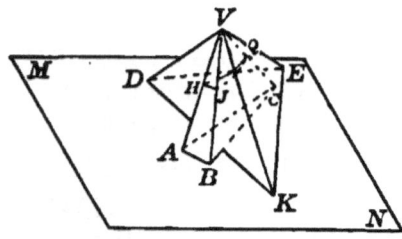

Fig. 259.

$\overline{VK}$ passing through the centre of the sphere, and being perpendicular to the plane AVC will pierce the surface of the sphere in the poles of the arc which make the intersection of the sphere with the plane AVC.

$\overline{VE}$ will pierce the surface in the poles of $\widehat{HJ}$; and $\overline{VD}$, in the poles of $\widehat{JQ}$.

These poles will be the vertices of a spherical triangle,* polar to the spherical triangle HJQ.

We have seen (§ 122) that if a spherical triangle is polar to a second spherical triangle, the second is polar to the first.

Hence: VA is perpendicular to the face KVE,

VB is perpendicular to the face DVE,

and VC is perpendicular to the face DVK.

The relations existing between the parts of supplementary triedrals are therefore mutual ones.

Applying further the relations established in § 122, we have the following

---

* Note. — This is omitted from the figure in order to save confusion.

THEOREM. *The facial angles of each of two supplementary triedrals are the supplements of the opposite diedrals of the other.*

128. Problems. — 1. Show that the sum of the facial angles of a polyedral angle is less than 360°.

Conceive a sphere of any radius having V for its centre and intersecting the polyhedral.

Then apply Ex. 11, § 124.

2. Show that the sum of the diedrals will be between the limits, $2n - 4$, and $2n$ rt. $\angle$, n being the number of faces.

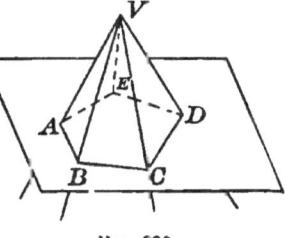

Fig. 290.

P

CHAPTER XII.

SURFACES.

129. Definitions. 1. If a portion of space be enclosed by planes, the portion enclosed is called a **polyedron**.

2. As has been seen, three planes, in general, intersect in a point and do not enclose any portion of space. It requires a fourth; although four planes may intersect so as *not* to enclose a portion of space.

They may all pass through one line, they may be parallel, or they may all pass through one point.

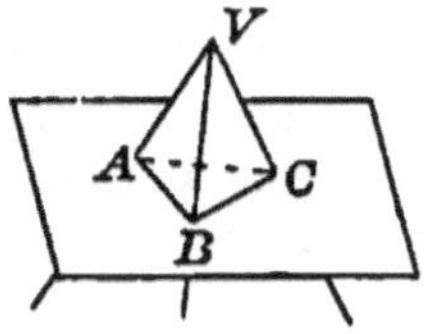

But in general they will enclose a portion of space, and the portion of space is called a **tetraedron**. The figure $V\text{-}ABC$ is a tetraedron. It has four triangular faces and four vertices (points through which three planes pass).

Fig. 291.

Any triangular face may be considered the base, and the vertex not in the plane of the base is called the **vertex** opposite the base.

3. If three planes intersect in parallel lines and the figure be intersected by two parallel planes, the enclosed portion of space is called a **triangular prism**.

4. If any number of planes intersect in parallel lines, and these parallel lines be inter-

Fig. 292.

sected by two parallel planes, the enclosed portion of space is called a **prism**.

The prism takes its name from the polygonal intersections of the parallel planes, called **bases**.

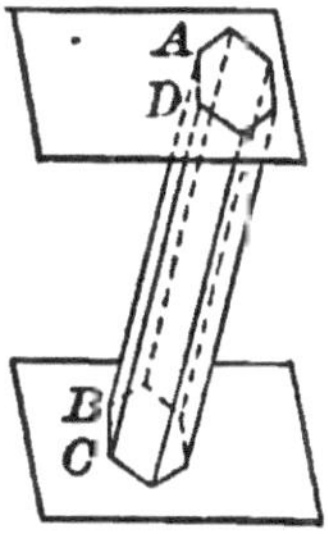

In the above figure the prism is hexagonal.

5. The parallel lines of intersection, as AB, are called **edges**.

6. Plane figures situated like $ABCD$ are called **lateral faces**. The student should prove that they are parallelograms.

Fig. 293.

7. If the edges are perpendicular to the bases, the prism is called **right**; if not perpendicular, the prism is called **oblique**.

A *right section* is made by passing a plane perpendicular to the edges.

8. The perpendicular distance between the parallel planes forming the bases is called the **altitude** of the prism.

9. If the secant planes are not parallel, the enclosed portion of space is called a **truncated prism**.

10. If the vertices of a plane polygon be joined to a point without the plane of the polygon, the figure thus determined is called a **pyramid**.

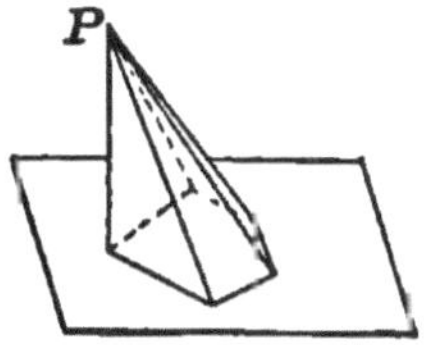

The polygon is called the **base**, and the given point, P, is called the **vertex**.

The triangles determined by the vertex and the sides of the polygon are called **lateral faces**; and the sides of the triangles not in the base are called **lateral edges**.

Fig. 294.

The altitude is the perpendicular distance of the vertex from the base.

11. Pyramids are classified in several ways:

(a) *Convex*, if the base is a convex polygon;

Concave, if the base is a re-entrant polygon.

(b) *Triangular, Quadrangular, Pentagonal*, etc., depending upon the number of angles in the base.

(c) *Regular*, if the base is a regular polygon and the projection of the vertex its centre. A *regular* pyramid is frequently called a *right* pyramid.

Irregular, if the base is not a regular polygon, or if the vertex is not projected in the centre of the base.

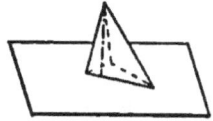

Fig. 295.

12. The altitude of a triangle forming one of the faces of a *regular* pyramid is called a slant height of the pyramid.

130. Theorem. *Sections made by parallel planes intersecting all the edges of a prism form equal polygons.*

Let $ABCDE$ and $FGHIJK$ be parallel plane sections. Then,

AB is parallel to FG (Prob. 1, § 114),

BC is parallel to GH (Prob. 1, § 114),

$\angle ABC = \angle FGH$ (Prob. 9, § 114).

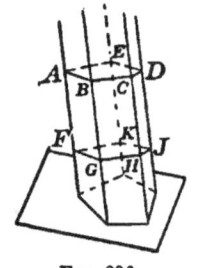

Fig. 296.

The same course being pursued about the perimeters of the polygons, they are found to have the sides and angles of one equal to the corresponding parts of the other. They are therefore equal.

Q. E. D.

A CYLINDRICAL SURFACE.

Exercises. — 1. Show that if a plane be passed parallel to the base of a pyramid, the intersection will be a polygon similar to the base.

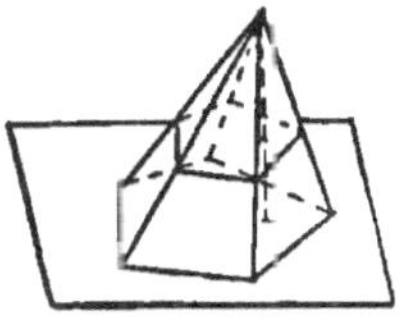

2. Show that the area of the section will be to the area of the base as the squares of the distances from the vertex.

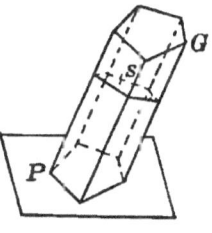

Fig. 297.

3. Show that if the section parallel to the base bisect the altitude, the area of the section will be one-fourth the area of the base.

4. Show that the lateral area of a prism equals the rectangle of the perimeter of a right section and a lateral edge.

131. Definitions. 1. If a straight line, not in the plane of a curve, be moved parallel to itself, and any point of the line move along the curve, a *cylindrical surface* will be generated.

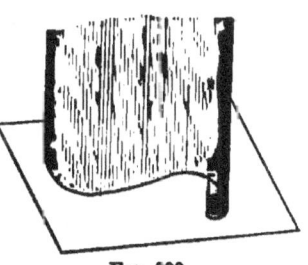

Fig. 298.

The straight line is called the **generatrix**, — or an element of the surface, — and the curve is called the **directrix**.

These terms are interchangeable; for the curve may be caused to move so that each point of it will generate a straight line parallel to a given straight line and will generate equal lengths in equal intervals.

Fig. 299.

2. Any section not parallel to the straight line elements, may be considered as the base. The character of the base, which may be any plane curve, closed or not, determines the general character of the cylindrical surface.

3. A **cylinder** is the definite portion of a cylindrical surface and volume, included between two parallel bases.

4. If the elements of the cylinder are perpendicular to the plane of the base, it is called a **right** cylinder.

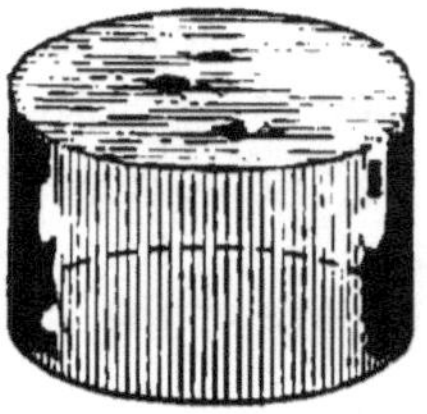

5. The cylinder most frequently considered is a right cylinder with a circular base.

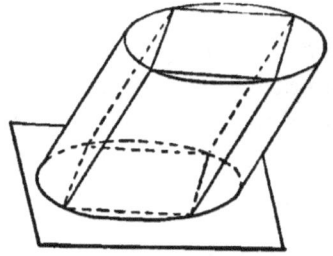

Fig. 300.

6. If the base of a cylinder be a closed curve, and if a polygon be inscribed within the base, and the elements of the cylinder through the vertices of this polygon be drawn, an inscribed prism will be formed; the volume of which will be less than the volume of the cylinder; and the area of the bases of which will be less than the area of the bases of the cylinder.

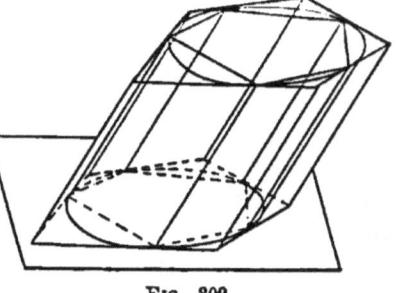

Fig. 301.

7. If under the same conditions as in Def. 6 a polygon be circumscribed about the base, and elements of the cylinder be drawn to the points of tangency, these elements and the sides of the polygon will determine planes which are said to be tangent to the cylinder. The number of such tangent planes

Fig. 302.

will equal the number of sides of the polygon, and the

PRISMS.

figure determined by them will be a prism circumscribing the cylinder.

This prism will exceed the cylinder in volume, and the area of its bases will be greater than the bases of the cylinder.

132. Definition. A **parallelopiped is a figure formed by six planes, parallel two and two.**

The lines of intersection bound the six faces.

A **rectangular parallelopiped is one whose angles are all right angles.**

Exercises.—1. Show that the opposite faces of a parallelopiped are equal parallelograms.

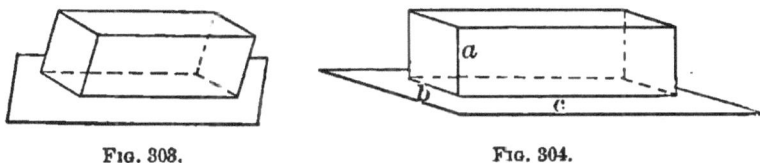

Fig. 303.   Fig. 304.

2. Show that the superficial area of a rectangular parallelopiped is $2(ab + ac + bc)$.

3. Show that the lateral area of a regular pyramid equals half of the product of the slant height by the perimeter of the base.

Fig. 305.   Fig. 306.

4. Show that to determine the lateral area of an oblique pyramid the area of each triangle will have to be determined separately.

5. Show that the cylinder about which prisms are circumscribed, or within which prisms are inscribed, will be the limit toward which these prisms approach, both in superficial area and in volume, as the number of lateral faces of the prisms is made to increase.

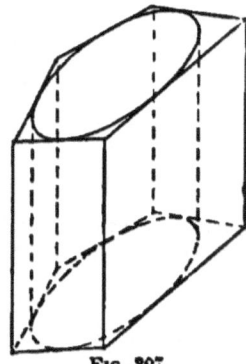

Fig. 307.

133. Definitions. We have in § 3 described a curve as generated by a point in motion.

If a point move on a curve from one position to another, it will have moved a *certain* distance.

If it move the same distance on a straight line, the portion of the straight line moved over is said to be the *rectilinear development* of the portion of the curve.

Illustration. — If a straight line be tangent to a circle at T, and the circle then be rolled upon the tangent until the point T again comes into the straight line at (T'),

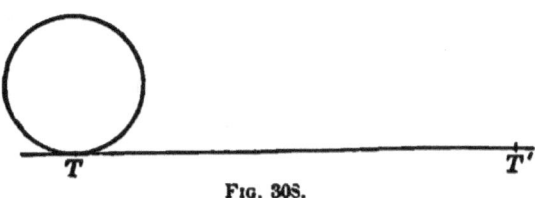

Fig. 308.

the segment of the straight line (TT') over which the circle has rolled is the rectilinear development of the circumference.

Theorem. *A surface that can be generated by a straight line moving parallel to itself, can be developed in a plane.*

CYLINDERS. 217

Let HK represent a cylinder, the elements of which are oblique to the parallel bases.

If through any point, P, of the surface, an element be drawn, and a plane be passed through the point perpendicular to the element, it will be perpendicular to all the

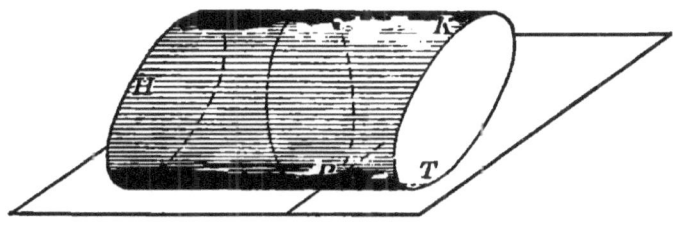

FIG. 809.

elements of the cylinder, will intersect the cylinder in a curve, and will be a *right* section.

If at P a tangent to the right section be drawn, the plane determined by this tangent and the element will be a tangent plane to the cylinder.

If the cylinder be rolled on the plane, the right section will be developed into a straight line in the tangent plane and every position of the generating element of the cylinder will come into the tangent plane, — being perpendicular to the right section and parallel to PT, after development as well as before.

When the original element of tangency shall have returned to the tangent plane, the cylinder will have made a complete roll, and every point and element of its surface will have come in contact with the plane.

The cylinder may therefore be developed on a plane.

134. Theorem. *Parallel plane sections of a cylinder are equal figures.*

A cylinder being generated by a straight line moving parallel to itself and passing through every point of a given curve, is such a surface that, being cut by parallel planes, the elements

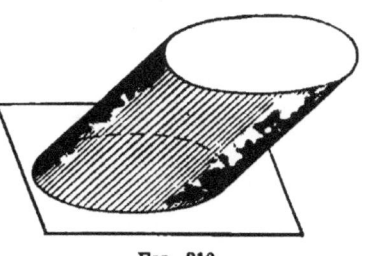

Fig. 310.

included between them will all be of equal length.

If one section be moved parallel to itself, and so that each point of the section shall follow the element passing through it, it will fall upon the other section and coincide with it point for point. Q. E. D.

Let the cylinder the bases of which are B and B' be a

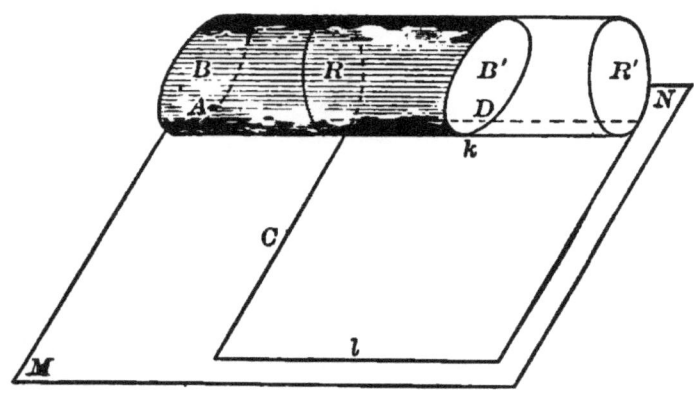

Fig. 311.

given cylinder with oblique bases, and MN a plane tangent along the element AD. If R be a right section and the portion between R and B be moved toward B' so that each point shall fall on the element passing through it, and each point be moved a distance equal to the portion of the elements between B and B', B will come to coincide with B', and R will have advanced to the position R'.

Thus it is seen that the convex area and the volume of the oblique cylinder are equivalent to the convex surface and the volume of the *right* cylinder RR'.

If this right cylinder be rolled upon the tangent plane until the element of initial contact shall again return to the plane, the surface of the cylinder will be developed on the tangent plane in the form of a rectangle, one side of which is the circumference of a right section, and the adjacent side is the portion of an element included between the bases.

Hence the following

THEOREM. *The area of the convex surface of a cylinder between parallel plane sections equals a rectangle, one side of which is the perimeter of a right section, and the adjacent sides the segment of an element included between the plane sections.*

Exercises. — 1. Find the convex (i.e. *curved*) surface of a right cylinder, the diameter of the base being 2 feet, and the altitude 100 feet.

2. How many square feet of sheet iron will be required to make 300 feet of 10-inch pipe, allowing one-sixth of the entire amount of sheet iron, for lapping and waste?

135. Definitions. If a straight line pass through a fixed point and follow a curve, it will in general describe a conical surface. The only exception is when the curve is a plane one and the fixed point is in the plane of the curve.

The point is called the **vertex** of the conical surface,

the curve, through the different points of which the straight line passes, is called the **directrix**, and the straight line is called an **element** of the surface. The parts of the cone separated from each other by the vertex are called **nappes**.

If the directrix be a circle and the vertex be in a perpendicular to the plane of the circle through its centre, the surface generated is called a **right circular cone**,* and the perpendicular is called the axis of the cone.

In all cones the surface is not limited in extent, but for certain purposes limited portions may be considered.

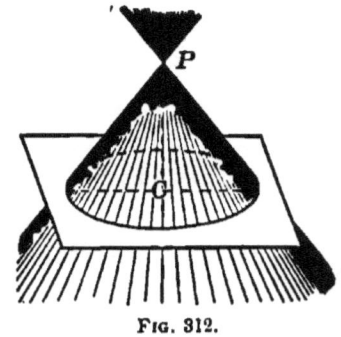
Fig. 312.

Any plane section may be taken as the base.

A **frustum** is the part of one *nappe* that lies between parallel planes.

If a straight line be drawn tangent to the directrix of a conical surface, the plane determined by that tangent and an element will not contain the adjacent elements, and is said to be tangent to the conical surface.

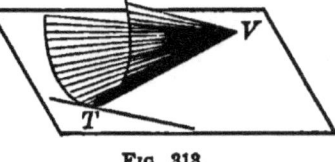
Fig. 313.

A tangent line to a curve may be rolled along on the curve. In each position it will determine with the ele-

---

* NOTE. — The right circular cone, when cut by planes in different positions, gives rise to the plane curves called conic sections, the most noted and most studied of all curves.

ment through the point of contact, a tangent plane. So that a tangent plane to a conical surface may be rolled upon the surface from any position to any other position.

Instead of rolling the plane on the surface, the surface may be rolled on the plane, and thus we see that any conical surface may be developed on a tangent plane; the directrix in general falling in a curve, and every point of the surface coming into the plane on which the development is made.

All tangent planes to a cone will pass through the vertex, and if one *nappe* is entirely on one side of the tangent plane, the other *nappe* will be on the other side.

Exercise. — Show that all plane sections parallel to the base of a right circular cone will be circles, the circumferences of which are to each other as their distances from the vertex; and the areas of which are to each other as the squares of these distances.

136. If at a point T of the base of a right-circular cone, a tangent to the circle be drawn, it will, with the element VT, determine a tangent plane MN.

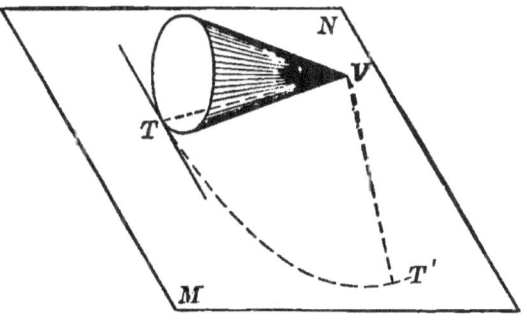

Fig. 314.

If the cone be rolled on the tangent plane until the element VT again comes into the plane, the portion of the surface between the base and the vertex will be developed into the sector of a circle, having the slant height of the cone for its radius and an arc equal to the circumference of the base.

Since by § 103, Ex. 1, we have the area of a sector equal to half the product of the arc and the radius, we have the

THEOREM. *The convex area of a right circular cone equals half the product of the slant height by the circumference of its base.*

Definitions. 1. If a polygon be circumscribed about the base of a cone and its vertices be joined with the vertex of the cone, a *circumscribed pyramid* will be formed.

2. If the vertices of an inscribed polygon be joined with the vertex of the cone, an *inscribed pyramid* will be formed.

Fig. 315.

Exercises. — 1. Show that the surface and volume of a cone are the limits toward which the surfaces and volumes of circumscribed and inscribed pyramids approach as the number of sides is increased.

2. Show the same for the volume and the lateral surface of a frustum.

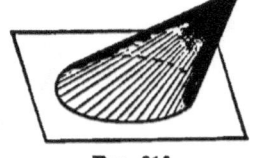

NOTE. — A cone not right is called oblique; and while its surface may be developed on a tangent plane, it will not be developed into a sector of a circle, and

Fig. 316.

the arc cannot in general be determined by the methods of elementary geometry.

137. Problems. — 1. Find an algebraic expression for the convex surface of a frustum of a right circular cone, the planes being passed perpendicular to the axis.

THE CONE.

Let C represent the circumference of the larger section and c represent the circumference of the smaller section.

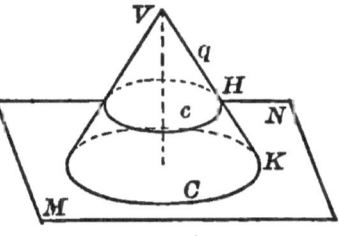

Fig. 317.

$s =$ slant height KH,
$q =$ slant height HV,
$L =$ lateral surface of VC,
$l =$ lateral surface of Vc,
$F =$ lateral surface of frustum Cc,

$F = L - l.$

$$L = \tfrac{1}{2} C (s + q), \qquad (\S 136)$$
$$l = \tfrac{1}{2} cq, \qquad (\S 136)$$

$$F = \tfrac{1}{2} C (s + q) - \tfrac{1}{2} cq$$
$$= \tfrac{1}{2} Cq - \tfrac{1}{2} cq + \tfrac{1}{2} Cs$$
$$= \tfrac{1}{2} (C - c) q + \tfrac{1}{2} Cs,$$

a form of expression for the required area, but one that involves q, a line which is not a part of the frustum.

In order to eliminate it from the expression, it is necessary to find relations between it and lines of the frustum.

$$\frac{C}{c} = \frac{s + q}{q},$$
$$Cq = cs + cq,$$
$$Cq - cq = cs,$$
$$(C - c) q = cs.$$

Substituting this value of $(C - c) q$ in the last obtained expression for F, we have,

$$F = \tfrac{1}{2} cs + \tfrac{1}{2} Cs \text{ or } F = \tfrac{1}{2} (C + c) s.$$
$$\therefore F = \tfrac{1}{2} (2 \pi R + 2 \pi r) s = \pi (r + R) s.$$

2. Show that $\tfrac{1}{2} (c + C)$ equals the circumference of a middle section, and that the area of the frustum may be described as the circumference of the middle section by the slant height:

$$F = 2 \pi \left(\frac{r + R}{2} \right) s.$$

3. Find an algebraic expression (formula) for the frustum of a cone in terms of the altitude EG.

From Prob. 2 we have:

$$F = 2\pi\, \overline{JM} \cdot \overline{BD}.$$

The *analysis* suggests the drawing of $DN \parallel$ to GE and $MQ \perp$ to BD; from which,

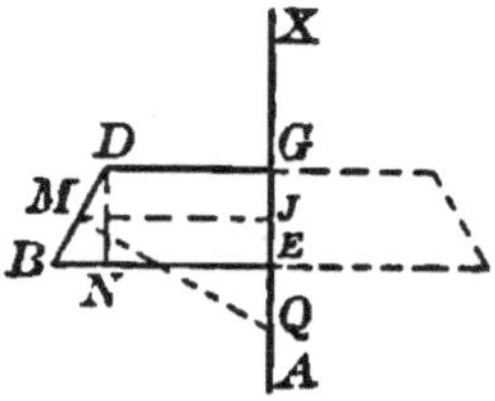

$$\frac{\overline{JM}}{\overline{QM}} = \frac{\overline{ND}}{\overline{BD}}$$

Fig. 318.

or $\overline{JM} \cdot \overline{BD} = \overline{QM} \cdot \overline{ND},$

$\therefore F = 2\pi\, \overline{QM} \cdot \overline{ND} = 2\pi\, \overline{QM} \cdot \overline{EG}.$ Q. E. F.

The expression last deduced shows that the area generated by revolving a segment of a straight line about another straight line in the same plane, will generate a surface which would equal the convex area of a cylinder having for its radius the perpendicular from the middle point of the generating line to the axis and for its altitude the projection of the given segment on the axis.

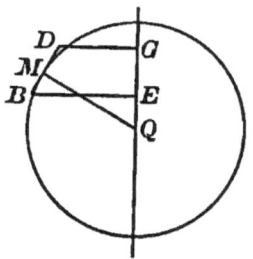

Fig. 319.

4. Show that if a regular polygon of an even number of sides be revolved about a diameter of the circumscribed circle, which is also a diagonal of the polygon, as an axis, the area generated will equal the convex surface of a right cylinder having for the radius of its base the apothem of the polygon, and the diameter of the circumscribed circle for its altitude.

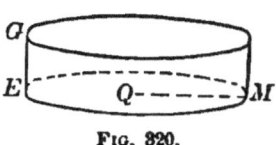

Fig. 320.

Or expressed differently: The area generated by the rotation of a regular polygon, as above described, will equal the convex area of a right cylinder, having for the diameter of the base the diam-

SPHERICAL SURFACE. 225

eter of the inscribed circle, and for its altitude the diameter of the circumscribed circle.

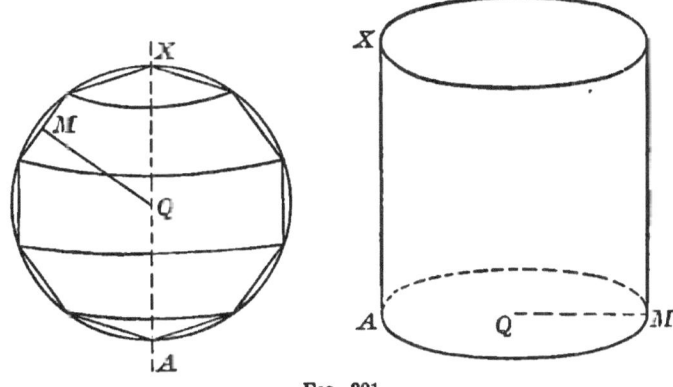

Fig. 321.

138. The surface of a cone or of a frustum of a cone may be considered as generated by a circle, the plane of which moves parallel to itself and the radius of which varies as its distance from a given point. For if we have a right circular cone or frustum, sections perpendicular to the axis will produce circles, the radii of which will be to each other as the distances of the sections from the vertex.

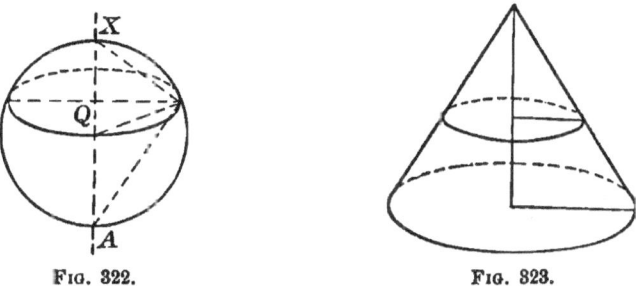

Fig. 322. Fig. 323.

The surface of a sphere or of a *zone* (§ 116) may be generated by moving a circle parallel to itself from A

toward X, remaining perpendicular to AX, and so changing the radius that its square shall equal the product of AQ and QX.

For if we have a sphere, the radius of any plane section will be a mean proportional between the segments into which the diameter, perpendicular to the section, is separated by it.

Theorem. *If a chord of a given circle be a tangent to a concentric circle, and the extremities of the chord be joined to the centre, and the figure thus determined be revolved about a non-intersecting diameter, the surface generated by the exterior arc will be greater than the surface generated by the chord, which will be greater than the surface generated by the interior intercepted arc.*

By § 103, $\widehat{BD} > \overline{BD} > \widehat{KE}$.

If *any* radius CP be drawn intersecting the inner circle at J, and the chord at T, P will be at a greater distance from the axis than T, and T will be at a greater distance than J. So that the circumference generated by P will be greater than that generated by T; which in turn will be greater than the circumference generated by J.

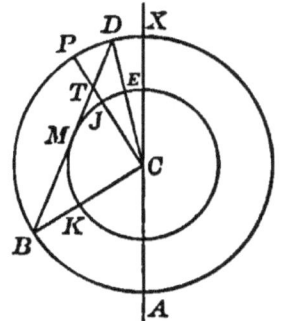

Fig. 324.

Revolved about AX, $\widehat{BD}$ will generate a zone, $\overline{BD}$ the frustum of a cone, and $\widehat{KE}$ a zone.

These may also be generated, as seen in the early part of this article, by circumferences, the planes of which are perpendicular to the axis.

SPHERICAL SURFACE.

In the cases now under consideration, the larger circumferences will be moved the greater distance and so will generate the larger area.

The zone formed by the revolution of $\overset{\frown}{BD}$ will be greater than the frustum formed by the revolution of $\overline{BD}$, and that in turn will be greater than the zone formed by the revolution of $\overset{\frown}{KE}$.

Hence the theorem is established.

139. Problem. *To find the expression for the surface of a sphere in terms of the square on its radius.*

As a consequence of the last article, if a regular polygon of an even number of sides, together with its circumscribed and inscribed circumferences, be revolved about a diagonal passing through a pair of opposite vertices, the surface generated by the polygon will be less than the surface of the circumscribed sphere and greater than the surface of the inscribed sphere.

In Prob. 4, § 137, it is shown that the area generated by the polygon equals the convex surface of a right cylinder, having for the diameter of its base the diameter of the inscribed circle, and for its altitude the diameter of the circumscribed circle.

If we should retain the same circumscribing circle and should double the number of sides of the polygon, the *diameter* of the inscribed circle would be increased, and the area generated by the polygon of the increased number of sides would be equivalent to a cylinder having the same *altitude* as before, but having a greater diameter of base.

Each time the number of sides of the polygon is doubled, the diameter of the cylinder having the equivalent convex area will be increased, but its altitude will not be changed.

We may conceive of this operation being performed as many times as we please, and at each one the surface generated by the polygon will be equivalent to the convex surface of a cylinder that remains constant in altitude, but the diameter of the base of which is increased.

This increase of diameter is limited, because the diameter of the inscribed circle will always be less than the diameter of the circumscribed circle. It will, however, approach the diameter of the circumscribed circle, as we conceive the number of sides of the polygon to be increased, and the difference between the two diameters may, by increasing the number of sides of the polygon, be reduced to as small a quantity as we please, and the two spherical surfaces be made to approach as near as we please.

As we perform these operations, or conceive them as being performed, there are two definite limits toward which the area generated by the polygon approaches:

These are the surface of the circumscribing sphere, and the convex surface of a cylinder having both diameter of base and altitude equal to the diameter of the circumscribing sphere.

Therefore, by the *limits axiom*, the surface of the circumscribing sphere will equal the convex surface of the cylinder having for its diameter and altitude the diameter of the circumscribing sphere.

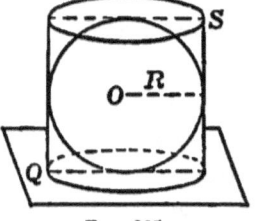

Fig. 325.

The convex surface of the cylinder is: $2\pi R \cdot 2R = 4\pi R^2$.

∴ Spherical surface $= 4\pi R^2$.

SPHERICAL SURFACE.

Note. — The surface of the sphere, the centre of which is O, equals the convex surface of the circumscribed cylinder; and equals four times the area of the base of the cylinder or the area of *four great circles*. Neither the convex surface of the cylinder nor the plane surface of four great circles can be applied to the surface of the sphere although they are equal.

Exercises. — 1. Show that the surfaces of any two spheres are to each other as the squares of their radii, or as the squares of any corresponding lines.

2. Find the surface of a sphere, the diameter of which is one and a half feet.

140. Problem. *To find the area of a zone.*

The arc BD, revolving about AK, will generate a zone.

The chord BD would generate the convex surface of a frustum, the area of which, as we have seen in Prob. 3, § 137, would equal the convex surface of a cylinder, having the apothem of BD for the radius of its base, and the projection QK for its altitude.

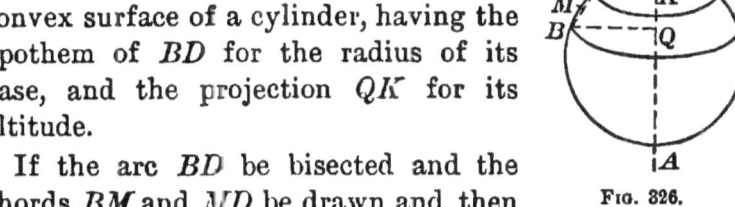

Fig. 326.

If the arc BD be bisected and the chords BM and MD be drawn and then revolved about AX, the surface will be the sum of two frustums, the combined convex area of which will be the convex surface of a cylinder having for the radius of its base the apothem of the new polygon, and for its altitude QK (a).

And further pursuing the matter in the same way as in the preceding article, we find the area of a zone equals the convex surface of a cylinder having the radius of the sphere for the radius of its base and the altitude of the zone for its altitude, or $Z = 2\pi R \cdot a$.

141. Definitions. A lune is the portion of the surface of a sphere included between two semicircles (§ 117).

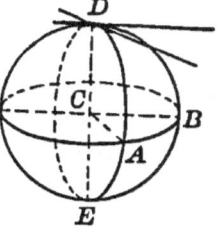

Fig. 327.

Two great circles of a sphere separate its surface into four lunes.

The angle of a lune is the angle that the planes of the great circles make with each other. As we have seen (§ 116), the angle between the planes is the same as the angle between the semicircles at their intersection, and that is measured by the arc of a great circle included between the arcs forming the lune, and 90° from the vertex.

The surface of a sphere may be considered as generated by the rotation of a semicircumference about its diameter as an axis.

Starting from its position DAE, the semicircumference may be conceived as rotating with uniform velocity until it shall have returned to its initial position. During this rotation the line CA rotates with uniform velocity and generates the angle which is the measure of the diedral.

Each starts at the same time, the motion is uniform, and the motion is completed at the same time.

When the semicircumference shall have arrived at the position DBE, the lune $A\text{-}DE\text{-}B$ will have been generated and it will be the same fractional part of the entire surface of the sphere that the angle ACB is of 360°.

$$\therefore \frac{\text{Lune}}{\text{Sphere}} = \frac{\text{Angle of lune}}{360°},$$

$$\text{or Lune} = \frac{\text{Angle of lune}}{360°}. \text{ Sphere.}$$

SPHERICAL SURFACE. 231

COROLLARY. *Two lunes are to each other as the angles of the lunes.*

$$\frac{L}{S} = \frac{\theta}{360°}, \quad \frac{l}{S} = \frac{\phi}{360°}.$$

$$\therefore \frac{L}{l} = \frac{\theta}{\phi}.$$

Exercise. — Show that the sum of a number of lunes equals a single lune, having for its angle the sum of the angles of the several lunes.

Remark. — A hemisphere may be regarded as a lune having an angle of 180°, and a sphere as a lune of 360°.

142. PROBLEM. *Deduce a formula* for the area of a spherical triangle.*

Let ABC represent any spherical triangle, one side,

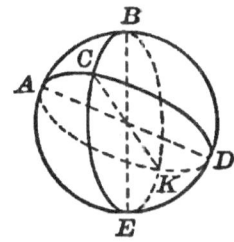

FIG. 328.

as a matter of convenience, being in the base of the hemisphere on which the spherical triangle is. (See note, § 121.)

The sp. $\triangle ABC$ + sp. $\triangle BCD$ = Lune $A°$,
  sp. $\triangle ABC$ + sp. $\triangle ACE$ = Lune $B°$,
  sp. $\triangle ABC$ + sp. $\triangle DEC$ = Lune $C°$,

3 sp. $\triangle ABC$ + sp. $\triangle BCD$ + sp. $\triangle ACE$ + sp. $\triangle BCE$ =
  Lune $A°$ + Lune $B°$ + Lune $C°$,

---

* A formula is the algebraic expression of a law.

or 2 sp. △ ABC + Hemi. = Lune $(A° + B° + C°)$,
   2 sp. △ ABC = Lune $(A° + B° + C°)$ − Hemi.,
   2 sp. △ ABC = Lune $(A° + B° + C° − 180°)$,
or sp. △ ABC = $\frac{\text{Lune}}{2}$ $(A° + B° + C° − 180°)$.

Exercises. — 1. Deduce a formula for the area of a trirectangular spherical triangle.

2. Deduce an expression for sp. △ ABC in terms of T (a trirectangular spherical triangle).

3. Deduce a formula for the area of a spherical triangle in terms of the sphere.

4. Deduce a formula for finding the area of any spherical *polygon, the angles* of which are given.

NOTE. — It has been shown that the sum of the angles of a spherical triangle is greater than 180°. The amount, in degrees, by which the sum of the angles exceeds 180°, is called the spherical excess $(A° + B° + C° − 180°)$.

143. THEOREM. *Similar surfaces are to each other as the squares of any homologous (corresponding) lines.*

Let P and p represent two similar polyedrons.

Analysis. — If the surfaces of these polyedrons are to each other as the square of any homologous lines, any corresponding portions of the surfaces would sustain the same relation.

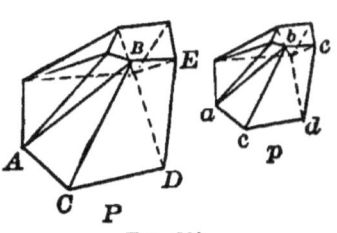

Fig. 829.

Demonstration. — The polyedra being similar, are bounded by similar plane polygons, similarly placed. Thus △ ABC and abc are similar and are placed with respect to the similar △ BCD and bcd so as to make the equal diedrals A–BC–D and a–bc–d.

The same may be said of any other adjacent faces.

By § 78 we have,
$$\frac{\triangle ABC}{\triangle abc} = \frac{\overline{AB}^2}{\overline{ab}^2} = \frac{\overline{BC}^2}{\overline{bc}^2} = \frac{\overline{AC}^2}{\overline{ac}^2},$$

$$\frac{\triangle BCD}{\triangle bcd} = \frac{\overline{BC}^2}{\overline{bc}^2} = \frac{\overline{BD}^2}{\overline{bd}^2} = \frac{\overline{CD}^2}{\overline{cd}^2}.$$

We have the common ratio $\dfrac{\overline{BC}^2}{\overline{bc}^2}$, so all the ratios are equal, and by § 79 we have,

$$\frac{\triangle ABC + \triangle BCD}{\triangle abc + \triangle bcd} = \frac{\overline{BC}^2}{\overline{bc}^2} = \frac{\overline{AB}^2}{\overline{ab}^2} = \frac{\overline{BD}^2}{\overline{bd}^2}, \text{etc.}$$

Again, $\quad \dfrac{\triangle BDE}{\triangle bde} = \dfrac{\overline{BE}^2}{\overline{be}^2} = \dfrac{\overline{BD}^2}{\overline{bd}^2},$ etc.

By § 79 we have,
$$\frac{\triangle ABC + \triangle BCD + \triangle BDE}{\triangle abc + \triangle bcd + \triangle bde} = \frac{\overline{BD}^2}{\overline{bd}^2} = \frac{\overline{AB}^2}{\overline{ab}^2}, \text{etc.}$$

In the same manner we may consider each of the corresponding polygons forming faces of the similar polyedra; and find that the ratios between the several similar polygons will be the same as the ratio of the *squares* of any homologous lines; and the sums of all of them *will have* the same ratio.

There are, as already seen, surfaces entirely curved, as a sphere, and surfaces partly curved and partly plane.

In any case the similar surfaces may be considered as the limits toward which similar inscribed or circumscribed polyedra approach as the number of faces is increased.

Lines may be chosen that shall be corresponding lines in the two similar surfaces under consideration and which shall be corresponding lines in the approaching polyedra.

The surfaces of the approaching similar polyedra are to each other as the squares of the homologous lines which may be so chosen as to remain unaltered during the approach.

$$\frac{S_1}{s_1} = \frac{L^2}{l^2},$$

$$\frac{S_2}{s_2} = \frac{L^2}{l^2},$$

$$\frac{S_3}{s_3} = \frac{L^2}{l^2}, \text{ etc.}$$

These surfaces represented by S_1, s_1, S_2, s_2, S_3, s_3, etc., which are similar at each step and continually approach the curved or mixed surfaces, always have the same ratio $\frac{L^2}{l^2}$. The limits toward which they approach will have the same ratio (§ 96). Hence the theorem.

NOTE.— In the above article is established the second of the three great principles of geometry as stated in § 69, Note.

Exercises. — 1. From the formula for the surface of a sphere, show that the surfaces of two spheres are to each other as the squares of their radii.

2. Show that the ratio of the areas of similar zones on different spheres equals the ratio of the squares on the diameters of the spheres.

3. Making use of a spherical blackboard, show how to find approximately the distance from New York to Queenstown.

4. With the same materials find approximately the distance from San Francisco to Yokohama, and show how near to the Aleutian Islands the shortest arc will pass.

CHAPTER XIII.

VOLUMES.

144. Definitions. 1. As has already been stated, a *volume* is an enclosed and limited portion of space.

The figure may be real or imaginary.

2. In § 57 the area of a rectangle is represented by ab, the product of two adjacent sides.

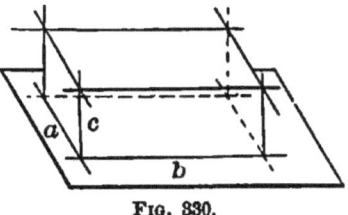

Fig. 330.

If that rectangle be moved perpendicularly a distance c, the volume generated is represented by abc. Expressed as a formula, it is, $V = abc$.

The figure is called a rectangular parallelopiped. It is also a right prism with rectangular base.

If the volume of one rectangular parallelopiped be represented by V, and the volume of another by v, the three edges meeting at a vertex in the one being a, b, and c, and in the other, x, y, and z, we have,

$$V = abc, \quad v = xyz.$$

By division, $\dfrac{V}{v} = \dfrac{abc}{xyz}$, ∴ the

THEOREM. *The volumes of two rectangular parallelopipeds are to each other as the products of the three edges meeting at a vertex.*

EQUIVALENT VOLUMES. 237

Exercises. — 1. Show that the theorem might be stated thus:
The volumes of two rectangular parallelopipeds are to each other as the products of their bases and altitudes.

2. If the bases happen to be equal, they are to each other as their altitudes.

3. If their altitudes happen to be equal, they are to each other as their bases.

4. Show that if the figures are similar, their volumes will be to each other as the cubes on any of the corresponding lines.

145. THEOREM. *An oblique prism is equivalent to a right prism, the base of which is a right section of the oblique prism, and the altitude of which is equal to an edge of the oblique prism.*

Let A–I represent the oblique prism, and K–N a right section, and QS another right section at a distance from K–N equal to an edge, AF, of the oblique prism.

Analysis. — *If* the right prism K–S be equivalent to the oblique prism A–I, the truncated prism K–I being common to both, the truncated prism F–S must be equivalent to the truncated prism A–N.

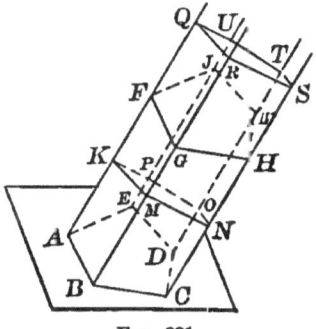

FIG. 331.

Demonstration. — Conceive the truncated prism A–N as moved in the direction of the edges. When it shall have been moved the distance AF, the oblique polygon $ABCDE$, forming its lower base, will coincide with $FGHIJ$, forming its upper base, the right section $KMNOP$ will coincide with the right section $QRSTU$, and the lateral faces will coincide, so that the oblique prism A-I will be converted, without change of volume, into the right prism K-S. Q. E. D.

146. Theorem. *An oblique parallelopiped is equivalent to a right parallelopiped having an equivalent base and the same altitude.*

Let $ABCDEFGH$ represent the oblique parallelopiped. The figure will be a prism, no matter which of the six faces be taken as a base.

If a plane section be made through $A \perp$ to AD and if a parallel plane be passed through D, we shall have (by § 145) the oblique prism converted into an equivalent right prism, $AIJKDONM$.

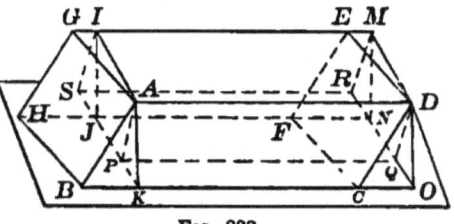

Fig. 332.

If $ADOK$ be considered as the base of the second prism, the edges parallel to AI will be oblique to it. A plane passed through AD and perpendicular to AI will make a right section, $APQD$. Another right section, $IMRS$, may be passed through IM, and we shall have (by § 145) the prism A-N, that is oblique with respect to the face A-O, converted into a right prism A-R, every angle of which is a right angle.

In these changes from one figure to an equivalent one, the altitude has remained the same, being the perpendicular distance between the planes H-C and G-D. By the first change the base H-C was converted into the rectangular base J-O, equivalent to H-C. And by the second change the rectangular base J-O was converted into the equivalent rectangular base S-Q.

These changes have been effected without change of volume. Q. E. D.

Exercise. — Show that the volume of any oblique parallelopiped is represented by the product of its base and altitude.

147. Problem. *To find an expression for the volume of a triangular prism.*

Let $ABCDEG$ represent an oblique triangular prism.

Analysis. — From what has preceded with regard to volumes, it seems probable that the volume of the oblique

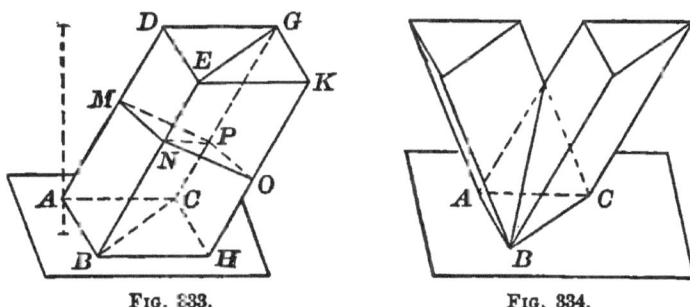

Fig. 333.　　　　Fig. 334.

triangular prism will be represented by the product of its base and altitude.

If it be so, we must arrive at the determination through the parallelopiped.

If
$BH \parallel$ to AC,
$CH \parallel$ to AB,
$EK \parallel$ to DG,
$GK \parallel$ to DE,

and HK be drawn, an oblique triangular prism will be annexed to the original triangular prism so as to form a parallelopiped, the volume of which is represented by the product of base and altitude.

The two triangular prisms have equivalent bases and

the same altitude, and each base is half the base of the parallelopiped.

The added prism we suspect is equivalent to the original prism, but it cannot be substituted for the original prism and be made to occupy the same space.

If we attempt the substitution, we shall have the base *BHC*, coinciding with the base *CAB*, by being reversed, but the elements will not take the same direction.

We may, however, convert the two oblique triangular prisms into equivalent *right* prisms having for their bases *right* sections, *MNP* and *NOP*, of the oblique prisms, and for their edges (altitudes) the edges of the oblique prisms. These right prisms may be substituted, the one for the other, and occupy the same space. But they are separately equivalent to the oblique triangular prism

Hence the oblique prisms are equivalent.

Parallelopiped $A\text{-}K = b \cdot h$.

Prism $A\text{-}G = \dfrac{b \cdot h}{2} = \dfrac{b}{2} \cdot h$.

But $\triangle ABC = \dfrac{b}{2}$.

∴ Prism $A\text{-}G = \triangle ABC \cdot h$, *i.e.* product of base and altitude.

Exercises. —1. Show that the volume of *any* prism may be represented by the product of its base and altitude.

2. Show that the volume of the prism $A\text{-}G$ may be represented by: $(\triangle MNP)(\overline{AD})$.

3. Show by the method of limits that the volume of any closed cylinder may be represented by the product of base and altitude.

4. Show that the diagonals of a parallelopiped mutually bisect each other.

VOLUMES OF PYRAMIDS. 241

5. Show that the volume of a triangular prism may be represented by half the product of one of its lateral faces by the perpendicular distance of the opposite edge from that face.

148. Theorem. *Triangular pyramids having equivalent bases and the same altitude are equivalent.**

Let A-BDC and N-OPQ represent the two pyramids.

If the altitudes be separated into any number of equal parts, and plane sections be passed through the points of division parallel to the bases, and auxiliary lines be drawn

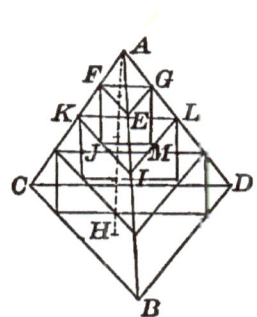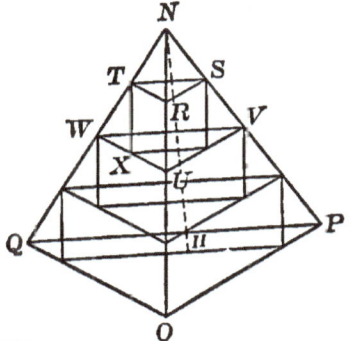

Fig. 335.

as indicated in the figures, a set of prisms will be formed, the number of which will be one less than the number of parts into which the altitude has been separated.

The plane sections at the same distances from the vertices will be equivalent.

$$\therefore EFG = RST,$$
$$IKL = VWU, \text{ etc.}$$

The three prisms in one figure will each be equal to one of the three prisms in the other figure; because prisms having equivalent bases and the same altitude are equivalent (§ 147).

---

*This Theorem is what is called a Lemma. It has been introduced to meet the requirements of the succeeding article.

Let Y_1 and Z_1 represent the sum of the three prisms inscribed in A-BCD and N-OPQ respectively.

Because each of the three prisms in one set is equivalent to one of the prisms in the other set,

$$\frac{Y_1}{Z_1} = 1.$$

If each of the previous divisions of the altitude be bisected and new prisms be constructed, with their edges parallel to AB and NO respectively, the two sets of prisms (Y_2 and Z_2) will be equivalent prism for prism.

$$\therefore \frac{Y_2}{Z_2} = 1.$$

With this and each further increase in the number of prisms (after the same manner) the volume occupied by the prisms will be increased; but at each step $\frac{Y_n}{Z_n} = 1$.

As the number of prisms increases in each pyramid, their sum approaches the volume of the pyramid as a limit, and $\frac{Y}{Z} = 1$, where Y and Z represent the volumes of the pyramids A-BCD and N-OPQ respectively.

Hence $Y = Z$. Q. E. D.

149. Problem. *Find an expression for the volume of a triangular pyramid.*

Let A-BCD represent the figure, the volume of which is to be determined.

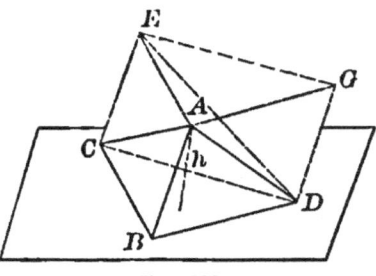

Fig. 836.

Analysis. — If a triangular prism as BCD-AEG be cut by the plane ACD, a triangular pyramid A-BCD and a quadrangular pyramid A-$ECDG$ will result. If

VOLUMES OF WEDGES. 243

the latter be cut by the plane AED, the triangular prism will be separated into three triangular pyramids.

The pyramids $A\text{-}EDG$ and $A\text{-}EDC$ have equivalent bases and the same altitude, and the pyramids $D\text{-}AEC$ and $D\text{-}ABC$ have equivalent bases and the same altitude.

If the pyramids having equivalent bases and the same altitude are equivalent in volume, each will be one-third of the volume of the prism.

Conclusion. — By § 148 $D\text{-}ABC = D\text{-}AEC = A\text{-}DEG.$

$\therefore A\text{-}BCD = \tfrac{1}{3} AEG\text{-}BCD,$

or $\qquad A\text{-}BCD = \tfrac{1}{3} h \cdot BCD = \tfrac{1}{3} h \cdot b.$ Q. E. F.

Exercises. — 1. Show that the volume of *any* pyramid may be represented by $\dfrac{b h}{3}$.

2. Show that the volume of a triangular pyramid and the volume of any pyramid may be represented by $\tfrac{1}{6} h\, (b + 4 m)$, in which b is the base and m is a section parallel to the base and midway between base and vertex.

3. $\qquad \dfrac{V}{v} = \dfrac{\dfrac{BH}{3}}{\dfrac{bh}{3}} = \dfrac{BH}{bh} = \left(\dfrac{B}{b}\right)\left(\dfrac{H}{h}\right).$

But $\qquad \dfrac{B}{b} = \dfrac{A^2}{a^2} = \dfrac{H^2}{h^2}.$

$\therefore \dfrac{V}{v} = \dfrac{H^3}{h^3} = \dfrac{A^3}{a^3},$ etc.

150. Definitions. Figures (1), (2), (3), and (4) represent

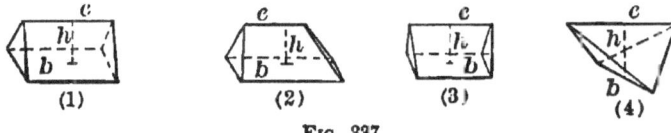

FIG. 337.

wedges. The *back* may be the same length as the *edge*, longer than the edge, shorter than the edge, or it may be

244 ELEMENTS OF GEOMETRY.

only a line. In the last case the wedge is a tetraedron, two opposite edges of which are considered back and edge, and the altitude of which is the perpendicular distance between back and edge.

Any of the wedges represented in (1), (2), and (3) may be converted into a wedge having the form of (4), and a quadrangular pyramid having its vertex at one extremity of the edge.

PROBLEM. *Find an expression for the volume of a wedge of form (4).*

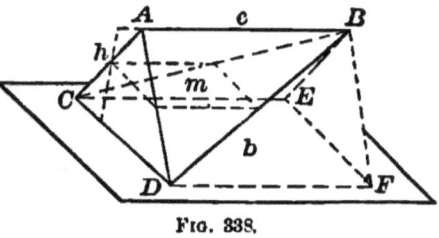

FIG. 338.

Let $ABCD$ represent the wedge, h being the perpendicular distance between the opposite edges AB and CD.

Through C and D, draw CE and DF parallel and equal to AB. Draw BE, BF, and EF.

The figure $ACDFEB$ is a triangular prism the volume of which by Ex. 5, § 147, is $\frac{1}{2} bh$.

A plane parallel to AB and CD, bisecting h, will intersect the wedge in a parallelogram (m), whose sides are one-half those of b.

Hence $m = \dfrac{b}{4}$, or $b = 4m$.

If W represent the volume of the wedge,

$$W = ACDFEB - (B\text{-}CDFE),$$
$$W = \frac{hb}{2} - \frac{hb}{3} = \frac{hb}{6} = \tfrac{1}{6} hb,$$

or $\qquad W = \tfrac{1}{6} h \cdot 4m.$* $\hfill$ Q. E. F.

---

*This form might be reduced to $\tfrac{2}{3} hm$, but this is not so convenient a form as the above, for purposes that will appear in the next article.

PRISMOIDAL FORMULA.

Exercise.—Show that the volume of each of the forms of wedges (1), (2), and (3), may be represented by:

$$\tfrac{1}{6} h(b + 4m),$$

in which b represents the area of the back, and m the area of the middle section.

151. Problem. *Deduce a formula for the volume of a figure, the upper and lower bases of which are plane polygons in parallel planes and the lateral faces of which are triangles or quadrangles.*

Let the accompanying figure represent a polyedron, the bases of which are any polygons whatever, in parallel planes, and the lateral faces triangles.

Some of the faces may be quadrangular plane figures, but these may always be separated into triangles, having their bases in the perimeter of either B or B'.

Such a figure may, by

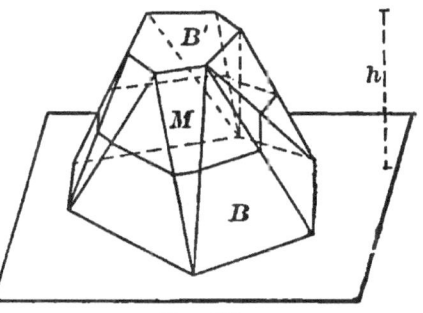

Fig. 339.

passing planes through selected lines and points, be cut into pyramids and wedges, the sum of the bases of which will be B and B', the sum of the middle sections, M, and the altitude of each sub-figure be h, the altitude of the entire figure.

The sum (P) of all the pyramids having their bases in B and their vertices in B', will give,

$$P = \tfrac{1}{6} h(B + 4m), \qquad \text{(Ex. 2, § 149)}$$

m representing the part of M which falls within these pyramids.

The sum (p) of all the pyramids having their bases in B' and their vertices in B, will give,

$$p = \tfrac{1}{6} h (B' + 4m'),$$

m' representing the part of M which falls within this second set of pyramids.

The remainder will be wedges of the form (4), § 150.

The sum (W) of these wedges, having lines in B' and B for their upper and lower bases, will give,

$$W = \tfrac{1}{6} h (4m'').$$

The total volume will be:

$$P = \tfrac{1}{6} h (B + 4m)$$
$$p = \tfrac{1}{6} h (B' + 4m')$$
$$W = \tfrac{1}{6} h (4m'')$$
$$\overline{V = \tfrac{1}{6} h (B + B' + 4M)} \qquad \text{Q. E. F.}$$

NOTE. — This formula, known as the prismoidal formula, is the greatest of all formulæ for the determination of volume.

Exercises. — 1. A *foot of lumber* is a piece a foot square and an inch thick. Find the number of feet of lumber in a telegraph pole 12 in. square at one end, 4 in. square at the other, and 20 ft. long.

2. Depending on § 136, Ex. 2, show that the prismoidal formula may be used to determine the volume of a frustum of a cone.

3. A tank, which is a frustum of a cone in shape, is 10 ft. deep, has a base diameter of 10 ft., and the diameter of its upper base is 8 ft. Find the number of gallons it will contain, each gallon being 231 cu. in.

152. PROBLEM. Show that a truncated triangular prism equals the sum of three pyramids, the altitudes of which will be the altitudes of the three vertices, and the bases of which will be the base of the truncated prism.

TRUNCATED TRIANGULAR PRISM.

The plane of A, E, and C determines a pyramid having

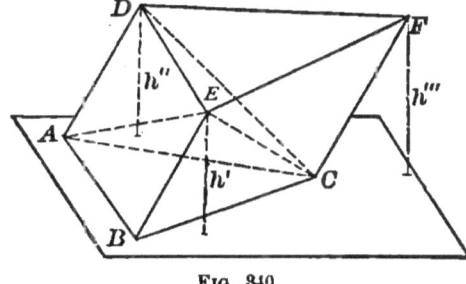

FIG. 340.

h' for its altitude and ABC for its base.

$$E\text{-}ABC = \frac{h'}{3} \cdot ABC.$$

The plane of E, D, and C determines the pyramids $E\text{-}ADC$, and $E\text{-}DCF$.

$$\frac{C\text{-}ADE}{C\text{-}ABE} = \frac{ADE}{ABE} = \frac{AD}{BE} = \frac{h''}{h'}.$$

$$\therefore C\text{-}ADE = \frac{h''}{h'}(C\text{-}ABE) = \frac{h''}{h'} \cdot \frac{h'}{3} \cdot ABC = \frac{h''}{3} \cdot ABC.$$

$$\frac{D\text{-}ECF}{A\text{-}EBC} = \frac{ECF}{EBC} = \frac{CF}{BE} = \frac{h'''}{h'}.$$

$$\therefore D\text{-}ECF = \frac{h'''}{h'}(A\text{-}EBC) = \frac{h'''}{h'} \cdot \frac{h'}{3} \cdot ABC = \frac{h'''}{3} \cdot ABC.$$

Q. E. D.

If V represent the volume of the truncated triangular prism,

$$V = \frac{h'}{3} \cdot ABC + \frac{h''}{3} \cdot ABC + \frac{h'''}{3} \cdot ABC.$$

$$V = ABC\left(\frac{h' + h'' + h'''}{3}\right).$$

PROBLEM. Show that the volume of a truncated quadrangular prism equals the product of its base by the average altitude of its vertices.

NOTE. — This fact is useful in computing earthwork.

Let A, B, C, and D represent points so located that their projection on a horizontal plane will form a square each side of which is, say, 10 ft. The elevation of the points A, B, C, and D is determined by taking their levels above some assumed plane, as MN.

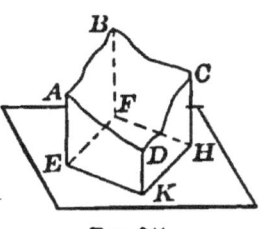

FIG. 341.

The formula determines the volume of the figure $ABCDEFHK$.

After the excavation, a resurvey and computation will determine the amount of dirt that has been removed.

153. PROBLEM. Show that the volume of a frustum of a triangular pyramid is equivalent to three pyramids each having the altitude of the frustum for its altitude, and having for their several bases the upper base, the lower base, and a mean proportional between the two.

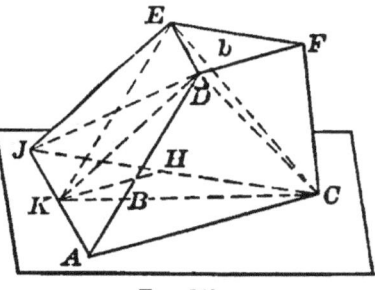

J, D, and C determine a plane which gives one of the pyramids, viz. D–JAC.

FIG. 342.

D, E, and C determine a plane which gives another of the pyramids, viz. C–DEF.

The remaining pyramid, D–EJC, is equivalent to K–EJC, which may be read, E–JKC. (DK was drawn parallel to EJ and KE and KC drawn.)

VOLUME OF A SPHERE.

If through the point K, KH be drawn parallel to AC,
$$\triangle JKH = \triangle EDF,$$
and (by § 80, Ex. 2) $\triangle JKC$ is a mean proportional between $\triangle JKH (= EDF)$ and $\triangle JAC$. Q. E. D.

Expressed as a formula,
$$V = \frac{Bh}{3} + \frac{bh}{3} + \frac{\sqrt{Bb} \cdot h}{3},$$
or
$$V = \frac{h}{3}(B + b + \sqrt{Bb}).$$

Exercises.—1. Show that for the frustum of *any* pyramid,
$$V = \frac{h}{3}(B + b + \sqrt{Bb}).$$

2. Show by the method of limits that the same formula may be applied to the determination of the volume of a cone frustum.

154. Problem. *Deduce a formula for the volume of a sphere.*

If a cube be circumscribed about a sphere, and lines be drawn from the vertices of the cube to the centre of the sphere, six pyramids will be formed, each with the same altitude (R), and the sum of the bases will be the surface of the cube. Representing the sum of the volumes of the pyramids by P_1, and the sum of the bases by S_1, we shall have,

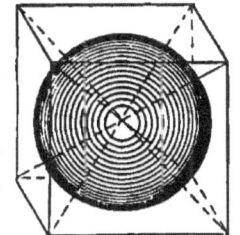

Fig. 343.

$$P_1 = \frac{RS_1}{3}, \text{ or } \frac{P_1}{S_1} = \frac{R}{3}.$$

If any number of planes be passed tangent to the sphere, they will cut off some of the volume of the cube exterior to the sphere.

If, then, lines be drawn from each of the vertices of the new circumscribing polyedron to the centre of the sphere, there will be formed as many pyramids as the polyedron has faces.

Representing the sum of the new pyramids by P_2, and the surface of the new polyedron by S_2, we will have,

$$P_2 = \frac{RS_2}{3}, \text{ or } \frac{P_2}{S_2} = \frac{R}{3}.$$

If, again, any number of planes be passed tangent to the sphere, we shall have at this stage of the process,

$$\frac{P_3}{S_3} = \frac{R}{3}.$$

With the next removal of exterior volume, the relations between the parts remaining will be,

$$\frac{P_4}{S_4} = \frac{R}{3}.$$

The same may be continued indefinitely, and each time we stop to consider the relations we shall have an equation of the form:

$$\frac{P_n}{S_n} = \frac{R}{3}.$$

The numerator of the first number continually diminishes and approaches the volume of the sphere as its limit; and the denominator approaches the surface of the sphere as its limit.

Hence if V represent the volume of the sphere and S its surface, we shall have for the relation between the limits (see § 96),

$$\frac{V}{S} = \frac{R}{3}; \quad \frac{V}{4\pi R^2} = \frac{R}{3},$$

or
$$V = \tfrac{4}{3}\pi R^3.$$

VOLUMES OF SIMILAR FIGURES. 251

Exercises. — 1. Apply the prismoidal formula to a sphere.

2. Apply the prismoidal formula to a hemisphere.

3. Show that the volumes of any two spheres are to each other as the cubes of their radii, or as the cubes of any corresponding lines.

4. The circumference of a great circle of a sphere being 6 ft., find the volume.

5. Find the relative volumes of a sphere and its circumscribed cylinder.

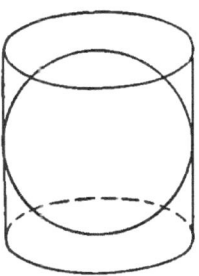

Fig. 844.

155. Theorem. *The volumes of similar figures are to each other as the cubes on any corresponding lines.*

Let P and p represent two polyedra that are similar, *i.e.* the figures are bounded by similar polygons which make diedrals with each other that in one are equal to those in the other.

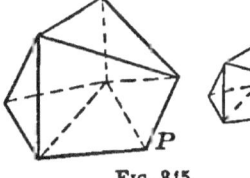

Fig. 845.

Analysis. — These similar polyedra may be separated into tetraedra that shall be similar, and *if* a relation between the volumes of similar tetraedra be known, the relation between the volumes of similar polyedra may be determined.

Demonstration. — (By § 149, Ex. 3.) Similar pyramids are to each other as the cubes on any corresponding lines. And if the pyramids be represented by P_1, p_1, P_2, p_2, etc., and the corresponding lines by L_1, l_1, L_2, l_2, etc., we will have,

$$\frac{P_1}{p_1} = \frac{L^3_1}{l^3_1}, \quad \frac{P_2}{p_2} = \frac{L^3_2}{l^3_2}, \text{ etc.}$$

But by reason of the fact that in similar figures corresponding lines are proportional,

$$\frac{L_1}{l_1} = \frac{L_2}{l_2} = \frac{L_3}{l_3}, \text{ etc.}, = \frac{L}{l}.$$

Recalling the fact that if proportions have common ratios they may be added term by term (see § 79),

$$\frac{P_1 + P_2 + P_3 + \text{etc.}}{p_1 + p_2 + p_3 + \text{etc.}} = \frac{P}{p} = \frac{L^3}{l^3},$$

which establishes the theorem as far as similar *polyedra* are concerned.

There are, however, volumes bounded by curved surfaces alone, and volumes bounded by surfaces some of which are curved and some of which are plane.

Such surfaces, provided they be similar, as stated in the announcement of the theorem, may have inscribed within them, or circumscribed about them, similar polyedra, which may be so constructed that they shall have at least one pair of corresponding lines which in the subsequent increase in the number of faces shall remain unchanged. These similar polyedra are to each other as the cubes on any similar lines.

If P_1 and p_1 represent, say, the circumscribing polyedra, and L and l a pair of corresponding lines that shall remain unchanged, we shall have,

$$\frac{P_1}{p_1} = \frac{L^3}{l^3}.$$

If the number of faces be increased by planes tangent to the volumes, we shall have,

$$\frac{P_2}{p_2} = \frac{L^3}{l^3}.$$

VOLUMES OF SIMILAR FIGURES. 253

At each step likewise,
$$\frac{P_n}{p_n} = \frac{L^3}{l^3};$$
and at each step we shall have P_n and p_n representing volumes which are diminishing and which are approaching fixed volumes as their limits.

By § 96 we arrive at the conclusion that
$$\frac{V}{v} = \frac{L^3}{l^3}. \qquad \text{Q. E. D.}$$

NOTE.—The principle just deduced is the third of the three great principles of elementary geometry heretofore alluded to.

Exercise.—Show how a numerical representative of a rectangular parallelopiped may be determined.

Suggestion.—Pursue the method of § 57.

CHAPTER XIV.

AN INTRODUCTION TO THE STUDY OF THE PLANE SECTIONS OF A RIGHT CIRCULAR CONE.

156. If we conceive of a plane tangent to a right circular cone (§ 136), and the plane through the element of tangency and the axis be represented by the page on which we have the figure, these planes will be perpendicular to each other (§§ 107, 112).

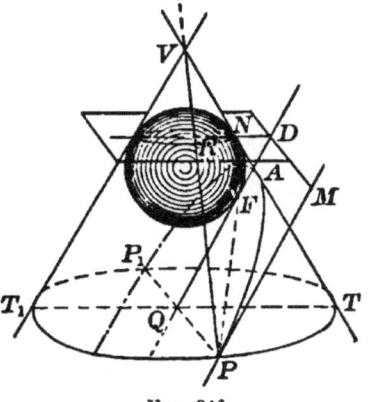

Fig. 346.

If a plane be passed parallel to the tangent plane, it will intersect one nappe of the cone and not the other. This line of intersection will be a plane curve which is called a **parabola**.

The curve will change its form depending upon the distance of the secant plane from the tangent plane.

There are many properties peculiar to the parabola, a few of which will be deduced in this chapter.

THEOREM. *The ratio of the distances of every point of a parabola from a certain point in its plane and from a certain straight line, also in the plane, equals* **1**.

THE PARABOLA.

Within the cone, tangent to the cone and also tangent to the secant plane, conceive a sphere to be constructed. It will be tangent to the cone on one of its small circles, the plane of which will be perpendicular to the axis of the cone; and for that reason, perpendicular to the plane on which the figure is represented.

The plane of the small circle, and the plane which intersects the surface of the cone in a parabola, are both perpendicular to the plane of the picture; and so (§ 111, Ex. 4) their line of intersection (DM) is perpendicular to the plane of the picture.

Let P be any point on the parabola. Join it to F (the point of tangency of the sphere and the secant plane), to V (the vertex of the cone), and draw $PM \perp$ to DM.

Through P pass a plane perpendicular to the axis of the cone; it will intersect the surface of the cone in the $\odot TPT_1$, and the plane of the picture in $\overline{TT_1}$.

The plane of the parabola intersects the plane of the picture in $\overline{QD}$.

$$\overline{PF} = \overline{PR}$$

(tangents to a sphere from P);

$$PR = TN$$

(portions of elements between parallel planes);

$$TN = TA + AN,$$
$$TA + AN = QA + AD$$

($\triangle AQT$ and AND are isosceles);

$$QA + AD = PM$$

(opposite sides of a rectangle are equal).

$$\therefore PF = PM,$$

or
$$\frac{PF}{PM} = 1.$$
Q. E. D.

256　ELEMENTS OF GEOMETRY.

NOTE. — Sections of a cone made by planes not parallel to a tangent plane will be considered after a few of the simpler properties of the parabola have been deduced.

Most of the properties of the Conic Sections are best developed by the methods of Analytic Geometry and Calculus; but because of their importance in Nature and in Art, a few of the simpler properties are here deduced.

157. Definitions. By reason of the property deduced in the preceding article, the parabola is often described as: The locus of a point moving in a plane so that the ratio of its distances from a fixed point and a fixed straight line equals 1.

The fixed point is called the **focus**.

The fixed straight line is called the **directrix**.

The $\perp FD$ is called the **axis**, because the curve is symmetrical with respect to this line: $P_1Q = PQ$, as may be show from the figure in § 156.

A is called the **vertex**. QP is called an **ordinate**.

The double ordinate through F is called the **parameter** or **latus rectum**.

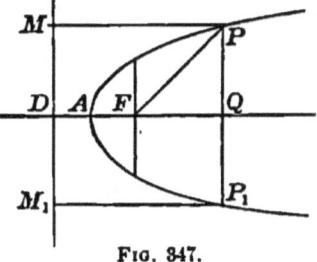

FIG. 347.

Exercise. — Show that the ordinate at the focus equals FD.

158. PROBLEM. *Having given the fixed point and the fixed line in a plane, to construct the parabola.*

Let AB be the given line and F the given point.

The analysis suggests the

Construction. — Draw a number of lines parallel to AB on the side that F is.

With F as a centre and the

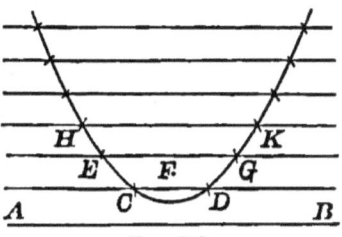

FIG. 348.

THE PARABOLA. 257

distance of the first line from AB as a radius, construct an arc, intersecting the first line.

These two points (C and D) of intersection are two points on the parabola.

Repeat the process, using the second line in the same way as the first was used, thus determining the points E and G.

In the same way H and K and any number of points may be determined.

Through these points draw a smooth curve; it will approximate closely to the required parabola.

A mechanical construction may be affected by using the edge of a ruler as the directrix, against which one perpendicular side of a right triangle is caused to slide, while against the other perpendicular side the marking point P presses a string, one end of which is secured at F and the other so fastened that

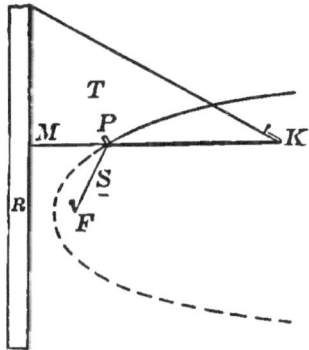

Fig. 349.

$$KP + PF = KM.$$

P will be a point on a parabola, because in any proper position
$$PF = PM.$$

Exercise. — Construct parabolas, having given the distance of F from the directrix, 1, 3, 10, and 20 centimetres.

159. Theorem. *If a line be drawn to the focus from the intersection of a secant with the directrix, it will bisect the exterior angle of the triangle formed by the chord and the focal radii.*

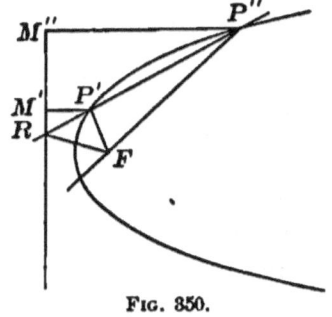

Fig. 350.

$P'M'$, and $P''M''$ being parallel,

$$\frac{P'M'}{P''M''} = \frac{RP'}{RP''}.$$

But $P'M' = FP'$,

and $P''M'' = FP''$.

$$\therefore \frac{FP'}{FP''} = \frac{RP'}{RP''}.$$

Therefore (§ 69, Prob. 7) the line FR bisects the exterior angle of the $\triangle P'FP''$.

Corollary. *If P' and P'' should coincide at P, PR would be a tangent and FR would be perpendicular to the focal chord through F and P.*

160. Theorem. *A tangent to a parabola bisects the angle formed by PF and PM.*

The analysis suggests that we establish the equality of the $\triangle MPR$ and FPR if possible.

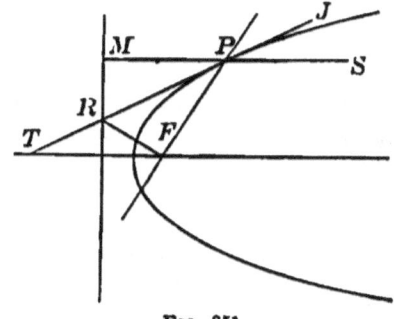

Fig. 351.

Demonstration. — These triangles are equal because each has a right angle; a side adjacent to the right angle in each equal; and the hypothenuse common.

$$\therefore \angle MPR = \angle FPR.$$

THE PARABOLA. 259

Note. — One of the important uses to which this property is put is the construction of parabolic reflectors.

$$\angle SPJ = \angle FPT.$$

And since the angle of incidence equals the angle of reflection, any ray of light emanating from F will be reflected to a line parallel to the axis.

The parabolic reflector is made by revolving a parabola on its axis. As a geometric figure it is called a paraboloid of revolution. If rays of light should enter the parabolic reflector parallel to the axis, they would all be reflected to the focus.

Exercises. — 1. Show that the distance from the focus to the point of tangency equals the distance from the focus to the foot of the tangent.

2. Show that if $\overline{FM}$ be drawn, it will be perpendicular to the tangent, and will be bisected by the tangent.

3. Show that the locus of the foot of a perpendicular from the focus to the tangent will be the tangent at the vertex of the curve.

4. Show that the projection on the axis of the segment of the tangent is bisected at the vertex. (This projection is called the *sub-tangent*.)

161. Definitions. A normal is a line perpendicular to a tangent and passing through the point of tangency.

A sub-normal is the projection on the axis of the segment of the normal included between the point of tangency and the axis. Thus QN is the sub-normal.

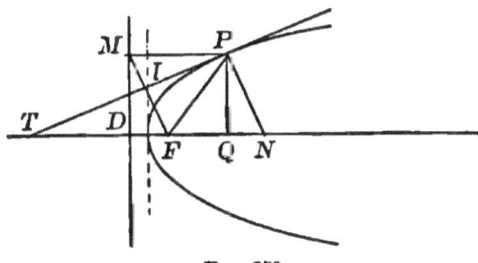

Fig. 352.

Theorem. *The sub-normal of a parabola is constant, and equals the distance of the focus from the directrix.*

$$\triangle PQN = \triangle MDF.$$
$$\therefore QN = DF.$$

Exercise. — 1. Show how, in three ways, to draw a tangent to a parabola at a given point of the curve.

162. Theorem. *If from the point of intersection of two tangents a straight line be drawn parallel to the axis of a parabola, it will bisect the chord joining the points of tangency.*

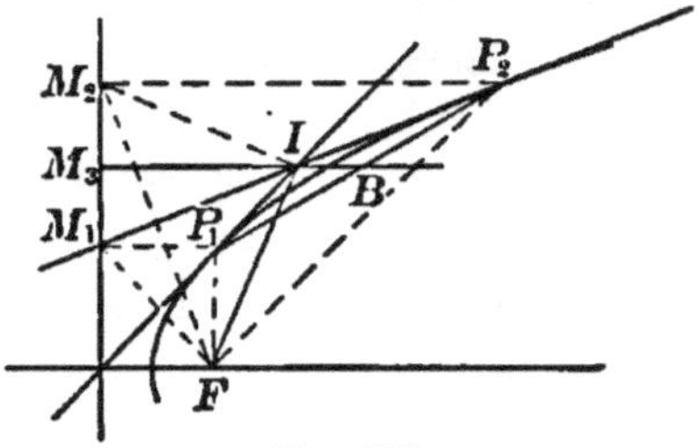

Analysis. — If $P_1B = BP_2$, M_1M_3 will equal M_3M_2 and the $\triangle M_1IM_2$ will be isosceles.

Demonstration. — Since IP_1 is the perpendicular bisector of FM_1 (Ex. 2, § 160), $IF = IM_1$.

Fig. 353.

Also since IP_2 is the perpendicular bisector of FM_2,
$$IF = IM_2.$$
$$\therefore IM_1 = IM_2,$$
and $\triangle M_1IM_2$ is isosceles.
$$\therefore M_1M_3 = M_3M_2$$
and $$P_1B = BP_2 \qquad \text{Q. E. D.}$$

Corollary. *If at the point of intersection of IB with the curve a tangent be drawn, it will be parallel to the chord P_1P_2, and the segment IB, which bisects the chord P_1P_2, is itself bisected by the curve.*

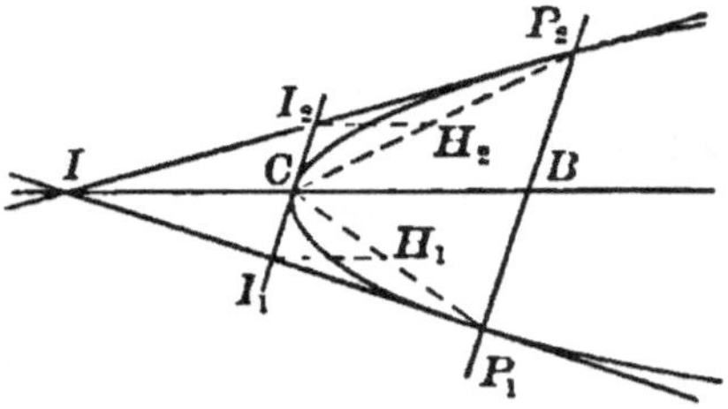

Analysis. — If a tangent at the point C is parallel to P_1P_2, and if it bisects IB, it will also bisect IP_1 and IP_2.

Demonstration. — Let

Fig. 354.

I_1I_2 represent the tangent at C, intersecting IP_1 and IP_2

THE PARABOLA. 261

By the *theorem*, a straight line through I_1, parallel to the axis, bisects CP_1.

I_1H_1 is parallel to IC (each being parallel to the axis).

∴ I_1 is the middle point of IP_1.

In the same manner it is shown that I_2 is the middle point of IP_2.

I_1I_2, bisecting two sides of the $\triangle IP_1P_2$, is parallel to the side P_1P_2 and bisects IB. Q. E. D.

PROBLEM. Show how to construct an arc of a parabola having given a pair of tangents and the points of tangency.

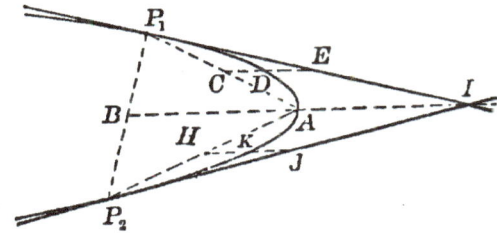

FIG. 355.

The analysis suggests the following:

Join P_1P_2; bisect it at B.
Join BI; bisect it at A.
Join AP_1; bisect it at C.
Draw CE parallel to BI; bisect it at D.
Join AP_2; bisect it at H.
Draw HJ parallel to BI; bisect it at K.

The points P_1, D, A, K, and P_2 are five points on the required curve.

As many points as are desired may be found in the same way, and the curve traced through them will be the arc of the parabola required.

NOTE. — The method described in the above problem is one of the processes frequently used for constructing parabolic railroad curves.

163. Problem. Show that the area included between an arc and a chord of a parabola is two-thirds the area of the triangle formed by the chord and the tangents at its extremities.

Let P_1P_2 represent the chord, P_1CP_2 the arc, IP_1 one tangent, and IP_2 the other.

B is the middle point of P_1P_2. Join IB. Draw a tangent at C and draw the chords CP_1 and CP_2.

The area of the $\triangle P_1CP_2$ is double the area of the $\triangle I_1II_2$; the base P_1P_2 is twice the base I_1I_2 and the altitudes are the same. The $\triangle P_1CP_2$ is said to be inscribed and the $\triangle I_1II_2$ is said to be escribed.

If at K and at H tangents be drawn to the curve, and chords be drawn from K to P_1 and to C, and also from H to C and P_2, we shall have an inscribed convex polygon, and an escribed re-entrant polygon. The additions to the area of the inscribed triangle (which have formed the inscribed polygon of 5 sides) are double the additions to the area of the escribed triangle (which have formed the escribed polygon of 5 sides).

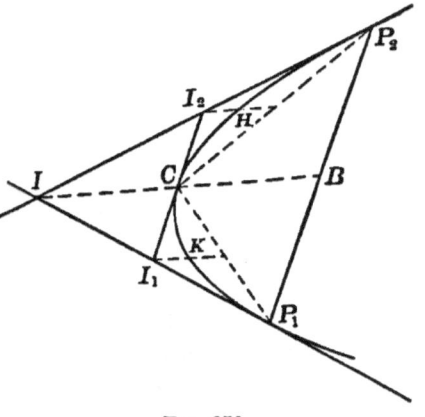

Fig. 356.

For these reasons the inscribed polygon of 5 sides has twice the area of the escribed polygon.

If at the points of intersection of the new tangents with those previously constructed, lines be drawn parallel to the axis, they will determine points on the curve, at which if tangents be drawn, and to which if cords be drawn from adjacent vertices previously determined, inscribed and escribed polygons of 9 sides would be formed, each having larger areas than the polygons of 5 sides. The increase in the inscribed polygon is double that in the escribed polygon.

The total area, then, of the inscribed polygon will be twice that of the escribed polygon.

Let $A_1, A_2, A_3, A_4, \cdots$ represent the areas of the inscribed polygons of 3, 5, 9, 17, etc., sides; let $a_1, a_2, a_3, a_4, \cdots$ represent the areas of the corresponding escribed polygons; and let A and a represent the limits toward which we approach as the number of sides is increased. The nature of the process and the steps already taken, show that

$$\frac{A_1}{a_1} = 2, \; \frac{A_2}{a_2} = 2, \; \frac{A_3}{a_3} = 2, \cdots \frac{A_n}{a_n} = 2.$$

The ratio as n increases is always 2; hence it will be 2 at the limit, i.e. $\frac{A}{a} = 2$, or

$$a = \frac{A}{2}.$$

At each step A_n approaches as its limit the area bounded by the first chord and its subtended arc; and a_n approaches as its limit the area bounded by the initial tangents and the given arc.

If T represents the area of the $\triangle P_1 I P_2$,

$$A + a = T, \text{ or } A + \frac{A}{2} = T;$$

$$\tfrac{3}{2} A = T, \text{ or } A = \tfrac{2}{3} T. \qquad \text{Q. E. F.}$$

Remark. — The area of the △ T equals a parallelogram having CB and BP_2 as its adjacent sides. If the chord be perpendicular to the axis, the parallelogram would be a rectangle.

It is interesting to note that we can convert an area bounded by a chord and an arc of a parabola into an equivalent square; a thing we are unable to do in the case of one bounded by a chord and a circular arc.

NOTE. — The parabola is one of the most interesting of curves.

The majority of comets move along parabolas with the sun at the focus.

The cables of a suspension bridge, when sustaining their own weight alone, form a curve called a *catenary*, but when loaded by a uniformly distributed weight (as when the road-bed is attached), the catenary is changed to a parabola.

A waterway in which the water will flow with uniform velocity, whatever its depth may be, will have a parabola for its cross-section.

The parabolic arch is the proper one to construct where a uniformly distributed load is to be sustained.

Both ends of a hen's egg are paraboloids of revolution.

The flight of a projectile approximates closely to a parabola.

PARTICULAR CASES.

164. If the plane which intersects the surface of a right circular cone in a parabola be moved parallel to itself until it coincides with the tangent plane to which it is parallel, the parabola will degenerate to a straight line, which is said to be a particular case of a parabola. If the plane be further moved in the same direction, the section will be a parabola in the other nappe.

If we conceive the elements of a right circular cone as increasing the angle which they make with a given base, the cone will approach a cylinder. When the angle that the elements make with the base shall equal 90°, the cone

THE ELLIPSE.

will have become a cylinder (a particular case of a cone), and a section made by a plane parallel to a tangent plane will be two parallel straight lines, another particular case of a parabola.

If we conceive the elements of a right circular cone as *decreasing* the angle which they make with a given base, the cone will approach the plane of its base, the parabola will approach a straight line; and when the cone shall have become a plane, the parabola will have become a straight line.

THE ELLIPSE.

165. If a plane be passed through a right circular cone intersecting all the elements in one nappe, the line of intersection is an ellipse.

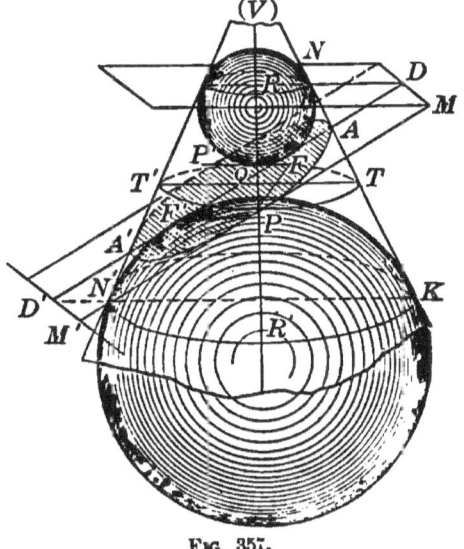

Fig. 357.

The section is a closed curve; for if a point on the section be moved along the curve in the same sense, it will return to its initial position.

Two spheres may be inscribed within the cone that shall at the same time be tangent to the secant plane (as at F and F'), and tangent to the surface of the cone (in circles, the planes of which are perpendicular to the axis of the cone).

Let the plane of the paper be the plane through the

axis of the cone and perpendicular to the secant plane. It will intersect the planes of the small circles of tangency in lines DM and $D'M'$, perpendicular to the plane of the picture.*

Let P represent *any* point of the section, and PQT a plane through P, perpendicular to the axis of the cone.

$$PF = PR = TN = TA + AN,$$
$$PM = QD = QA + AD.$$

From the similar $\triangle$ AQT and ADN, we have,

$$\frac{TA}{AN} = \frac{QA}{AD}; \quad \frac{TA+AN}{AN} = \frac{QA+AD}{AD};$$

$$\frac{TA+AN}{QA+AD} = \frac{AN}{AD}$$

$$\therefore \frac{PF}{PM} = \frac{AN}{AD} = \frac{AF}{AD}.$$

$\frac{AN}{AD}$ is constant for this position of the secant plane and is < 1, because in the $\triangle$ AND, AN is opposite a smaller angle than AD is. Hence we have the

Theorem. *In the plane of an ellipse there is a point (called a focus), and a straight line (called a directrix), the distances from which to any point of the curve will have a fixed ratio, and that ratio will be less than* 1.

Exercises. — 1. Show that $\frac{PF'}{PM'} = \frac{A'N'}{A'D'}$.

2. Show that $\frac{A'N'}{A'D'} = \frac{AN}{AD}$, and hence $\frac{PF}{PM} = \frac{PF'}{PM'}$ (§ 65, Ex. 2).

3. Show that $PQ = QP'$.

---

*It is expressly understood that the sketches which serve as figures for purposes of determining relations between parts are not perspective drawings.

THE ELLIPSE.

166. Theorem. *The sum of the distances of any point of an ellipse from two certain fixed points in its plane is constant.*

By the figure of the preceding article, it is seen that

$$PF = PR \text{ and } PF' = R'P.$$
$$\therefore PF + PF' = R'R \text{ (a constant).} \quad \text{Q. E. D.}$$

Problems. — 1. Show that $AF = A'F'$.

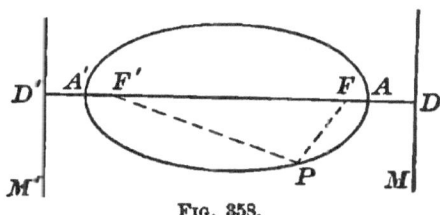

FIG. 358.

Since $PF + PF'$ is a constant, the sum will not be altered by taking P in a particular position. If it be moved to A and then to A', we shall have,

$$AF + AF' = A'F' + A'F,$$
or $$AF + AF + FF' = A'F' + A'F' + FF',$$
or $$2\, AF = 2\, A'F',$$
$$\therefore \quad AF = A'F'.$$

2. Show that $PF + PF' = AA'$.

Definitions. — The points F and F' are called the *foci*.

The lines DM and $D'M'$ are called *directrices*.

A figure is symmetrical with respect to a line if a perpendicular to the line at any point intersects the figure at equal distances on either side of the line. The perpendicular bisector of AA' is called the *minor axis*.

The segment AA' through the foci is called the *major axis* and is a line of symmetry.

An ellipse may be defined as the *locus* of a point moving so that the ratio of its distances from a fixed point and from a fixed straight line is constant and less than 1.

An ellipse may also be defined as the *locus* of a point moving so that the sum of its distances from two fixed points is constant.

Exercises.—1. Assume a point and a straight line and construct by points an ellipse having the ratio $\dfrac{PF}{PM} = \dfrac{3}{5}$.

2. Two points are 10 centimetres apart. Construct by continuous motion an ellipse the major axis of which is 12 centimetres.

3. From any point within an ellipse the sum of the distances to the foci is less than the major axis, and from any point without an ellipse the sum of the distances to the foci is greater than the major axis.

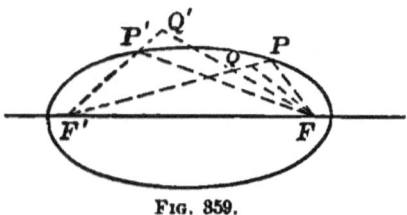

Fig. 359.

Note.—There are so many ways of constructing an ellipse with reasonable accuracy that there is no excuse for making up a combination of arcs of circles and calling the figure an ellipse.

167. Theorem. *If a straight line be drawn through a point of an ellipse so as to make equal angles with the focal radii to the same point, it will be a tangent to the curve.*

A tangent is the limiting position toward which a secant approaches as the two points of intersection approach coincidence. When the secant shall have become a tangent, all of its points except the point of contact will lie without the curve.

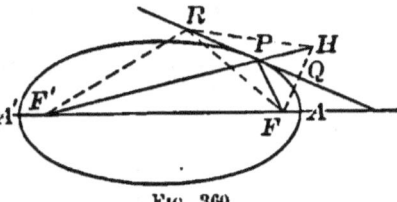

Fig. 360.

THE ELLIPSE.

Produce $F'P$ so that PH shall equal PF; making PHF an isosceles triangle. Through P draw $RQ \perp$ to FH; it will make equal angles with PF and PF'.

If RQ be a tangent, every point except the point of contact P must be shown to be exterior to the ellipse.

Let R represent any point other than P.

$$F'R + RH > F'H. \qquad (\S\ 167,\ \text{Ex. 3})$$

But $RH = RF$ and $F'H = AA'$.

$$\therefore F'R + RF > AA'.$$

Hence the point R, which may be any point on the line other than P, is exterior to the curve, and RQ is a tangent.

Note. — If an ellipse be revolved about its major axis, a *prolate ellipsoid* of revolution is generated. If revolved about its minor axis, an *oblate ellipsoid* of revolution is formed. The earth is approximately an oblate ellipsoid of revolution.

168. Theorem I. *An ellipse is determinable when its axes are given.*

We have seen that an ellipse is the locus of a point which moves so that the sum of its distances from F and F' equals AA' ($2\ a$).

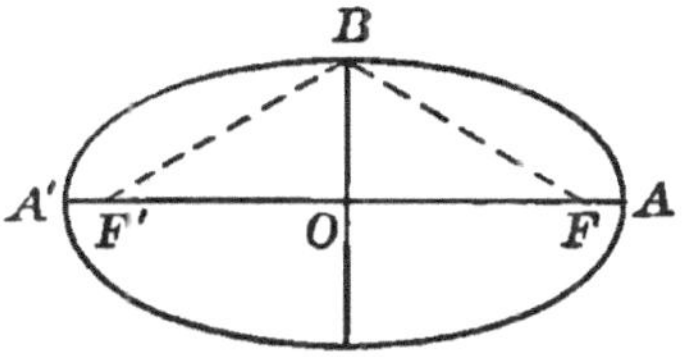

When the moving point is at B,
$$BF = BF' = a.$$

Fig. 361.

If, then, the segments which are to be axes are placed so as to mutually bisect each other at right angles, the foci may be determined by constructing an arc of a circle with B as centre and a as radius. The foci and the major axis determine the ellipse (§ 166); therefore the axes do.

270 ELEMENTS OF GEOMETRY.

THEOREM II. *A section oblique to the elements of a right circular cylinder will be an ellipse.*

Let APA' represent the oblique section. Within the cylinder and tangent to the secant plane, conceive spheres to be inscribed as indicated in the figure; they will be tangent to the secant plane at two points, as F' and F''.

Draw PF, PF', and RR'.

$$PF = PR,$$
$$PF' = PR',$$
$$PF = PF' = RR' \text{ (a constant).}$$

∴ the section is an ellipse. Q. E. D.

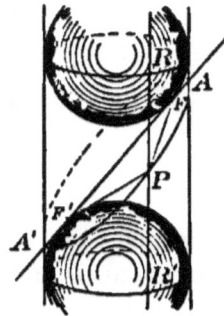

FIG. 362.

COROLLARY. *Any right section is a circle; so that an ellipse may be projected into a circle.*

169. PROBLEM. *If a and b represent the semi-axes of an ellipse, show that the area is represented by πab.*

Let APA' represent an oblique section of a right circular cylinder and KHK' a right section of the same passed through the middle point of AA'.

BB' will be perpendicular to AA' and will be the minor axis of the ellipse, as well as being the diameter of the ⊙ KHK'.

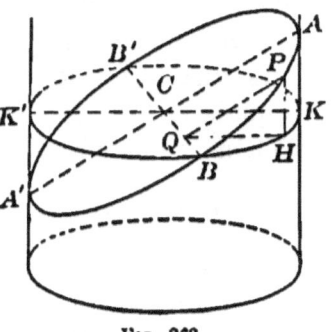

FIG. 363.

From P draw $PQ \perp$ to CB and PH perpendicular to the plane of the circle.

QH will also be perpendicular to CB (§ 112).

THE ELLIPSE.

The △ PQH and ACK are similar.

$$\frac{QP}{QH} = \frac{CA}{CK} = \frac{a}{b}.$$

If the point P should move from B to A to A', the segment PQ remaining perpendicular to BB' would generate half the area of the ellipse; and the segment HQ would generate half the area of the circle.

Each line segment would remain perpendicular to BB', would move the same distance, and the ratio of their lengths would remain the same, therefore

$$\frac{\text{Half the Ellipse}}{\text{Half the Circle}} = \frac{a}{b},$$

or

$$\frac{\text{Ellipse}}{\text{Circle}} = \frac{a}{b}$$

$$\frac{\text{Ellipse}}{\pi b^2} = \frac{a}{b}$$

$$\text{Ellipse} = \frac{\pi a b^2}{b} = \pi ab. \qquad \text{Q. E. F.}$$

Exercises. — 1. Deduce the same expression for the area of an ellipse by inscribing a polygon within the ellipse, comparing its area with the area of the projected polygon in the circle, and then by the theory of limits determine the relation between the areas of the ellipse and the circle.

2. Show that an ellipse has a centre, *i.e.* a point through which, if chords be drawn, they will be bisected.

3. Show that if a line be drawn from the centre of the ellipse to the point Q, in the figure of § 167, its length will be a. Hence the locus of the foot of a perpendicular from a focus to a tangent is the circumference of a circle on the major axis as a diameter.

This circle is called the director circle.

NOTE. — The earth's meridians are ellipses, the axis of the earth being the minor axis of each ellipse.

It is this fact which gives rise to the statement that the Mississippi River runs up hill. Its mouth is further from the centre of the earth than its source.

The locus of the earth in its annual motion about the sun is approximately an ellipse with the sun at one focus.

If a light (as an electric arc-light) were placed at one focus of a prolate ellipsoid, all rays reflected from the surface would meet at the other focus. (Established by § 167.)

Whispering galleries are also made which depend upon this principle.

PARTICULAR CASES.

170. If a plane which intersects the surface of a cone in an ellipse be moved parallel to itself until it passes through the vertex, the ellipse will degenerate to a point. If the plane be moved further, an ellipse in the other nappe will be the result.

If a plane which cuts an ellipse from a right circular cone be rotated about some axis toward the position of being perpendicular to the axis, the foci will approach each other, and when the plane becomes perpendicular to the axis, the foci and centre will coincide and the ellipse will become a circle.

THE HYPERBOLA.

171. Definitions. If a plane intersect a right circular cone and make with the plane of a circular section an angle greater than that made by the elements of the cone with the same section, it will intersect both nappes of the cone.

The line of intersection is called a **hyperbola**. In general it is in two branches, neither of which close.

THE HYPERBOLA. 273

The two branches are, however, superimposable, as will be shown later in this chapter.

P, P', P''' represent three points of the secant plane, S and S' represent spheres tangent to the cone and to the secant plane.

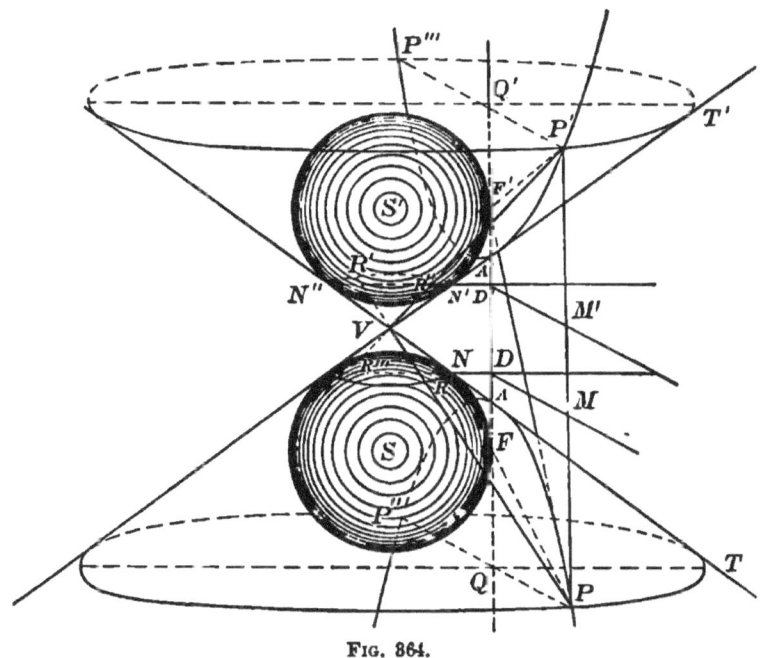

FIG. 864.

Let P be any point in one branch. Through P pass a plane perpendicular to the axis of the cone, giving the circle PTP''.

PV is an element of the cone, tangent to the spheres at R and R'.

PF and PF' are straight lines to the points of tangency of the inscribed tangent spheres.

T

DM and $D'M'$ are the intersections of the secant plane with the planes of the circles of tangency of the spheres and the cone.

QQ' is the intersection of the secant plane with the plane of the picture.

$$PF = PR = TN = TA + AN,$$

$$PM = QD = QA + AD.$$

$$\therefore \frac{PF}{PM} = \frac{TA + AN}{QA + AD}.$$

But $\triangle AQT$ and AND are similar.

$$\therefore \frac{TA}{AN} = \frac{QA}{AD}, \text{ or } \frac{TA + AN}{AN} = \frac{QA + AD}{AD};$$

or
$$\frac{TA + AN}{QA + AD} = \frac{AN}{AD}.$$

$$\therefore \frac{PF}{PM} = \frac{AN}{AD}.$$

For this particular section, AN and AD are fixed; and AN, being opposite a greater angle than AD in the $\triangle AND$, is greater than AD.

Hence $\dfrac{PF}{PM}$ is constant and >1.

Note. — This ratio, which for the *Parabola* $=1$, for the *Ellipse* <1, and for the *Hyperbola* >1, is sometimes called the *Boscorich ratio*, but is generally known as the *eccentricity* and is represented by e.

THE HYPERBOLA. 275

172. Problems. — 1. To find an expression for the ratio $\dfrac{PF'}{PM'}$ and compare it with $\dfrac{PF}{PM}$.

$$\frac{PF'}{PM'} = \frac{PR'}{QD'} = \frac{TN''}{QD'} = \frac{TA + AN''}{QA + AD'}.$$

But $\quad \dfrac{TA}{AN''} = \dfrac{QA}{AD'}$, or $\dfrac{TA + AN''}{AN''} = \dfrac{QA + AD'}{AD'}$:

or $\quad \dfrac{TA + AN''}{QA + AD'} = \dfrac{AN''}{AD'} = \dfrac{AN}{AD} = \dfrac{PF}{PM}$.

$\therefore \dfrac{PF'}{PM'} = \dfrac{PF}{PM}$.

2. To find an expression for the ratio $\dfrac{P'F'}{P'M'}$ and compare it with $\dfrac{PF}{PM}$.

$$\frac{P'F'}{P'M'} = \frac{P'R''}{Q'D'} = \frac{T'N'}{Q'D'} = \frac{T'A' + A'N'}{Q'A' + A'D'}.$$

But $\quad \dfrac{T'A'}{A'N'} = \dfrac{Q'A'}{A'D'}$, or $\dfrac{T'A' + A'N'}{A'N'} = \dfrac{Q'A' + A'D'}{A'D'}$,

or $\quad \dfrac{T'A' + A'N'}{Q'A' + A'D'} = \dfrac{A'N'}{A'D'} = \dfrac{AN}{AD}$ (§ 65, Ex. 2) $= \dfrac{PF}{PM}$.

Hence $\dfrac{P'F'}{P'M'} = \dfrac{PF}{PM}$.

3. Draw $P'F$ and show that

$$\frac{P'F}{P'M} = \frac{P'F'}{P'M'} = \frac{PF'}{PM'} = \frac{PF}{PM} = e.$$

4. Show by Fig. 362 that $PF' - PF$ is constant, and that $P'F - P'F'$ equals the same constant.

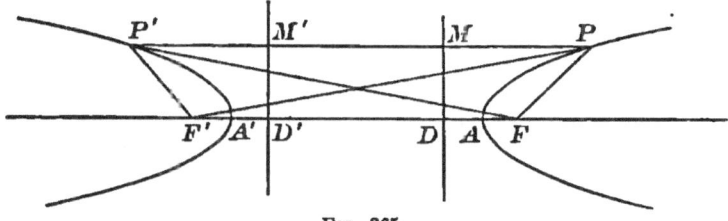

Fig. 365.

F and F' are called foci, and DM and $D'M'$ are called directrices.

276 ELEMENTS OF GEOMETRY.

5. Find a line that shall represent the constant, $PF' - PF$.

A and A' being points of the curve,

$$AF' - AF = K$$
$$A'F - A'F' = K$$
$$\overline{AA' + A'F' - AF = K}$$
$$AA' + AF - A'F' = K$$ } or $\quad 2\,A'F' - 2\,AF = 0,$
$\qquad\qquad\qquad\qquad\qquad\qquad\quad A'F' = AF.$

$$\overline{2\,AA' = 2\,K}$$

or $\quad K = AA'.$

In addition to determining the fact that $PF' - PF = AA'$, we have determined that $A'F' = AF$.

6. Show that $A'D' = AD$, and that the hyperbola is symmetrical with respect to FF' and also with respect to the perpendicular bisector of FF'.

NOTE.—Because of the relation developed in Prob. 4, the hyperbola may be defined as: The locus of a point moving in a plane so that the difference of its distances from two fixed points, also in the plane, will remain fixed.

If the condition that the locus be a plane curve be removed, the locus would be a hyperboloid of revolution of two nappes. If revolved on the vertical axis of symmetry, a single sheet will be generated, which would, upon investigation, turn out to be a warped surface as well as a surface of revolution, and one that could be generated by a straight line revolving about another straight line, not in the same plane with it.

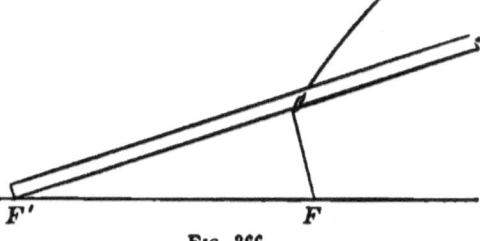

FIG. 366.

7. Construct a plane curve such that the ratio of the distances of all points of the curve from a fixed point and from a fixed straight line shall be $\tfrac{2}{3}$.

8. Show how to construct a hyperbola by a continuous motion.

173. Theorem I. *If any point (H) be on the concave side of a branch of a hyperbola, i.e. within the cone, $HF' - HF > AA'$.*

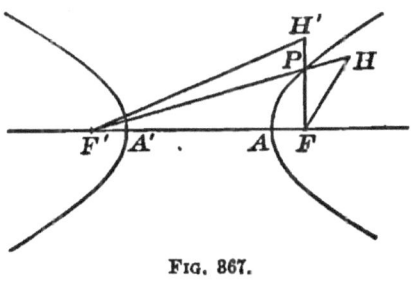

Fig. 867.

$$PF' - PF = AA$$
$$PH = PH$$
$$\overline{PF' + PH - PF = AA' + PH}$$
$$FH - PF < PH$$
$$\overline{\therefore HF' - FH > AA'} \qquad \text{Q. E. D.}$$

Theorem II. *If any point (H) be on the convex sides of both branches of a hyperbola, i.e. not within the cone, $H'F' - H'F < AA'$.*

$$PF' - PF = AA'$$
$$PH' = PH'$$
$$\overline{PF' - PH' - PF = AA' - PH'}$$
$$H'F' - PF' < PH'$$
$$\overline{H'F' - PH' - PF < AA'}$$
$$H'F' - (PH' + PF) < AA'.$$
$$\therefore H'F' - H'F < AA'. \qquad \text{Q. E. D.}$$

Theorem III. *If a straight line be drawn through a point of a hyperbola so as to make equal angles with the focal radii to the same point, it will be a tangent to the curve.*

Analysis. — If the straight line TP touch the hyperbola at P, and every other point of TP be exterior to the curve, it will be a tangent.

Demonstration. — The line TP, by hypothésis, bisects the $\angle F'PF$.

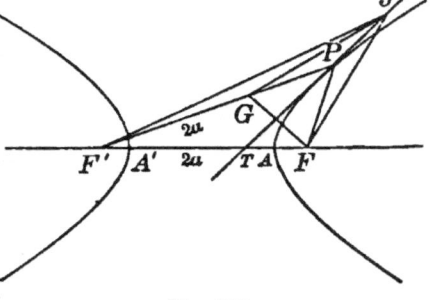

Fig. 368.

If FG be drawn perpendicular to TP, GPF will be isosceles, TP will be the perpendicular bisector of FG, and since $PF' - PF = AA'$, $PF' - PG = AA'$ or $F'G = AA'(2a)$.

If any point of the angle bisector, as J, be joined with F, G, and F', we will have,
$$JF' - JF = JF' - JG < 2a.$$

Hence J (any point other than P) is exterior to the curve and TP is a tangent at P. Q. E. D.

Remark. — A *normal* would bisect the adjacent angle formed by the focal radii to the point of tangency.

Note. — In the parabola where there is but one focus, the other may be said to be on the axis at infinity. This view makes possible the general enunciation. Lines drawn from the foci of a conic section to any point of the curve make equal angles with the tangent at that point.

THE HYPERBOLA.

174. Since the tangent at P bisects the $\angle F''PF$, it separates $F''F$ into segments proportional to the adjacent sides. $\therefore F'T$ is the greater segment and the foot of a tangent will always fall within $A'A$.

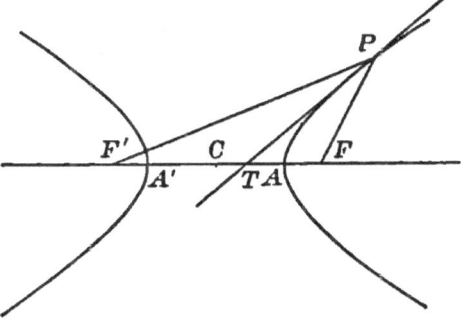

Fig. 369.

As the point of tangency (P) becomes more remote from A, the point T approaches C, but does not reach it until P shall have passed to infinity; in which case $F'P$, TP, and FP will be parallel. In this limiting position the tangent is called an **asymptote**. It is a determined line toward which the curve approaches, and to which it is tangent at an infinite (∞) distance from C.

In §173 it is shown that a perpendicular from F to a tangent will, if produced its length, have its extremity in the line drawn from F'' to the point of tangency.

Because $F'P$ and CP are parallel, the $\angle F''GF$ will be 90° and its vertex must be in the circumference of a circle on FF'' as a diameter.

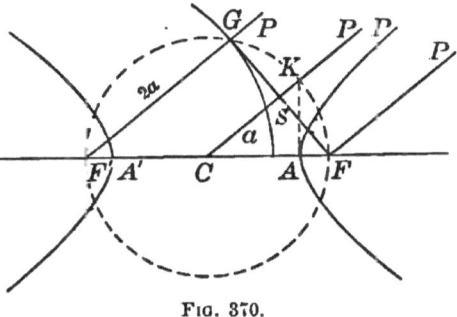

Fig. 370.

It was also shown that $F''G = 2a$; and G must be in

the circumference, having F'' for its centre and $2a$ for its radius.

Hence, to construct an asymptote:

On FF'' as a diameter, construct a circumference.

With F'' as a centre and AA' as a radius, construct an arc intersecting the preceding one at G and G'.

Through C draw lines parallel to $F''G$ and to $F''G'$. They will be the asymptotes.

If the construction be made from F instead of from F', it would be found that the asymptotes to one branch are also asymptotes to the other.

If at A a perpendicular be erected, $AK = FS$, $CK = CF$. If CA be represented by a, CF by c, and AK by b, we have $c^2 = a^2 + b^2$.

The asymptotes are the diagonals of a rectangle, the sides of which are $2a$ and $2b$.

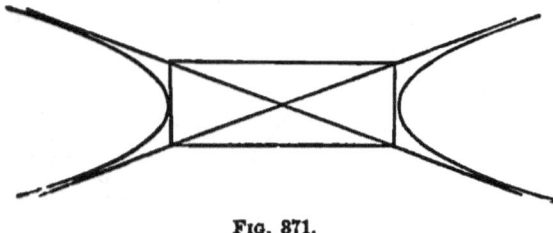

FIG. 871.

Exercises.—1. Show how to draw a tangent to the hyperbola at a given point on the curve.

2. Show how to draw a tangent to a hyperbola from any point. What limitations are there as to the position of the point from which the tangent is to be drawn? Show that in general there may be two tangents.

PARTICULAR CASES.

175. If the plane which cuts a hyperbola from a cone be moved parallel to itself and *toward* the vertex, the vertices of the curve and the foci approach each other, and when the plane passes through the vertex of the cone, the vertices and foci will all coincide and the hyperbola will become two intersecting straight lines.

If the plane be moved parallel to itself and *away* from the vertex of the cone, the vertices and the foci of the curve will recede from each other, the curvature in the neighbourhood of the vertices will decrease, and the hyperbola will approach two parallel lines, both at infinity and at an infinite distance from each other.

If the cone with a given circular base should approach a cylinder, the section which yields a hyperbola will approach parallel straight lines.

If the angle which the elements make with the axis should approach 90°, the vertices of the two branches would approach each other, the curvature would diminish, and when the angle should become 90°, the cone would form a plane and the two branches of the hyperbola would fall together in a single straight line.

NOTE. — Recalling the figures wherein the sections of a cone are represented, and conceiving a perpendicular to the plane of the picture as erected at A, we may by passing a plane through this perpendicular, and rotating it about the perpendicular as an axis, obtain the different sections in turn.

Beginning in such a position that the section shall be a hyperbola, and rotating toward the position which will give a parabola, we see that the vertex and the focus of the second branch recede from the first, and that when the plane becomes parallel to a tangent plane, the second vertex and its adjacent focus will have passed to infinity.

The instant that the rotation carries the secant plane beyond the position when it gives a parabola, we get an ellipse with the second focus and vertex at a very great distance from the first; as the rotation continues, the foci approach each other, and when the rotating plane is perpendicular to the axis the foci will have coincided and the section will be a circle, *i.e.* a particular case of an ellipse.

If the rotation be continued, the foci will again separate and the ellipse become narrower and narrower until the secant plane becomes a tangent plane, when the ellipse will have degenerated to a right line, which is also a particular case of the parabola.

This last position might have been reached by rotating the secant plane the reverse way, in which case the straight line would be the limit toward which the hyperbola would approach.

The properties of the ellipse and hyperbola are complementary and may be deduced together.

One conic section may be projected into another by considering some point not in the plane of the section as the source of light, and the shadow be intersected by planes occupying different positions.

SOME PROBLEMS IN SOLID AND SPHERICAL GEOMETRY.

1. Show that if a straight line be perpendicular to a plane its projection on any other plane will be perpendicular to the line of intersection of the two planes.

2. Show how to draw a straight line through the vertex of a triedral so as to separate it into three isosceles triedrals.

3. Show how to circumscribe a circle about a spherical triangle.

4. Show how to inscribe a circle within a spherical triangle.

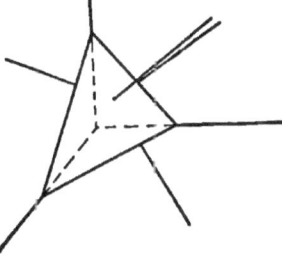

FIG. 372.

5. Through a point on the surface of a sphere to draw an arc of a great circle that shall be tangent to a given small circle.

6. Use concentric spheres to show that the intensity of a light varies inversely as the square of the distance from the source.

7. The three angles of a spherical triangle are 64°, 85°, and 122°, on a sphere the diameter of which is 18 inches; find the area.

8. Regarding the earth as a sphere the radius of which is 3956 miles, find the area of a trirectangular spherical triangle.

9. Assume numerical values for each of the parts which are sufficient to determine a spherical triangle and make the construction on a spherical blackboard.

10. A sphere, the radius of which is 1, is circumscribed by a cube, a cylinder, and a cone, the elements of which make an angle of 30° with the axis; find the relative surfaces and volumes.

11. A sphere is circumscribed by a cylinder; the lower base of the cylinder is the base of a right circular cone, the vertex of which is at the centre of the upper base; find the volume which is common to the cone and the sphere.

12. To find *approximately* the distance of the horizon on the ocean as seen from an elevation.

$$e(e + 2R) = d^2.$$

For any elevations that exist on the earth, the e that is added to $2R$ may be neglected without serious error.

The e which is a *factor cannot* be neglected.

$$\therefore e = \frac{d^2}{2R} \text{ or } d^2 = 2eR.$$

If R be 4000 miles,

$$e = \tfrac{1}{8000} d^2 \text{ (in miles)}.$$

In feet,

$$e = \frac{d^2}{8000} \times 5280 = \frac{528}{800} d^2 = \frac{66}{100} d^2 = \frac{2}{3} d^2.$$

$$\therefore e \text{ (in feet)} = \tfrac{2}{3} d^2 \text{ in miles}.$$

If d be one mile, $e = \tfrac{2}{3}$ of a foot, or 8 inches.

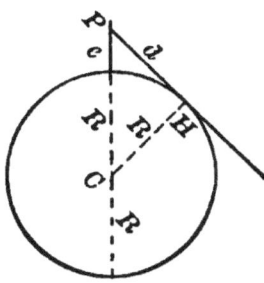

Fig. 373.

13. Mt. Washington is 6288 feet high; how far is the ocean horizon from it?

14. Standing on the deck of a ship is a man whose eyes are 20 feet from the water. A second ship, the hull of which stands 12 feet out of water, is seen to be just "hull down." What is the distance apart of the two ships?

15. Two triedrals have two facial angles of one equal to two facial angles of the other, but the enclosed diedrals are not equal; what relations exist between the third facial angles in each?

16. Show that the three planes which bisect the diedrals of a triedral, pass through one line.

17. Show that the three planes passed through the bisectors of the facial angles, perpendicular to those faces, pass through one line.

PROBLEMS IN LOCI.

18. Find the locus of points in space which are equally distant from two given points.

19. Find the locus of points which are equally distant from three given points.

20. Find the locus of points which are equally distant from two given planes.

21. Find the locus of points which are equally distant from three given planes.

22. Find the locus of a point which moves so that its distances from the three edges of a triedral are always equal to each other.

23. Find the locus of points in a given plane which are equally distant from two points which may or may not be in the given plane.

24. Find the locus of a point which moves so that the ratio of its distances from two parallel lines always equals 1.

25. Find the locus of a point which moves so that the ratio of its distances from a fixed line and a fixed point of that line is constant.

26. Find the locus of points which are the centres of the sections of a sphere made by planes. (*a*) Which pass through a fixed line. (*b*) Which pass through a fixed point.

27. Find the locus of the centre of a sphere which is tangent to three given planes.

SOME PROBLEMS INVOLVING THE INTERSECTION OF LOCI.

28. Find the locus of a point which moves so that the ratio of its distances from two points equals 1; and the ratio of its distances from two other points equals 1.

29. Find the locus of points which are at a given distance from P, and twice as far from Q.

30. Find the locus of points which are at a given distance, and the ratio of the distances of which from K and Q equals 1.

31. Locate a point which shall be equally distant from two given planes and from a given point.

32. Locate a point, the ratio of the distances of which from two planes equals 1, and the ratio of its distances from two points equals 1.

33. Find the locus of a point which moves so that it remains at a fixed distance from a given straight line, and the ratio of its distances from two fixed points equals 1.

34. Locate a point so that the difference of the squares of its distances from two given points is constant.

35. Find the locus of points which are equally distant from the surface of a sphere, and which are so situated that the ratio of their distances from two given points is 1.

36. Locate a point in a plane so that the sum of the squares of its distances from two given points not in the plane, is fixed; and its distance from a given line of the plane is also fixed.

37. Find the locus of a point from which two spheres subtend equal visual cones, and a given segment of a straight line subtends a given angle.

THE FIVE REGULAR POLYEDRONS.

38. Show that *four* equal equilateral triangles may be placed so as to enclose a volume.

Definition.—The figure thus formed is called a regular *tetraedron*.

39. One edge of a tetraedron is a; deduce formulæ for its superficial area and for its volume.

Fig. 874.

40. Show that six equal squares may be placed so as to enclose a volume.

Definition.—The figure thus formed is called a regular *Hexaedron* or Cube.

Remark.—The superficial area and the volume we are already familiar with.

PROBLEMS.   287

NOTE. — It is recommended to the student that he make pasteboard models of the regular polyedra.

41. Show that *eight* equal equilateral triangles may be so placed as to enclose a volume.

Definition. — The figure thus formed is called a regular *octaedron*.

FIG. 375.

42. One edge of a regular octaedron is a; deduce formulæ for the superficial area and the volume.

43. Show that *twelve* regular pentagons may be so placed as to enclose a volume.

Definition. — The figure thus formed is called a regular *dodecaedron*.

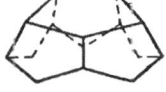

FIG. 376.

44. One edge of a dodecaedron is a; deduce formulæ for the superficial area and the volume.

Remark. — The student who solves this problem without assistance is presumed to have a pretty good working power in geometry.

45. Show that *twenty* equal equilateral triangles may be so placed as to enclose a volume.

Definition. — The figure thus formed is called a regular *icosaedron*.

46. One edge of a regular icosaedron is a; deduce formulæ for the superficial area and the volume.

47. Show that the *five* regular polyedra already named are the only regular polyedra that can be formed.

NOTE. — There are many combinations of polygons that may be used to enclose a volume; as one may see by looking over the sketches of any work on crystallography.

NUMERICAL EXERCISES.

48. An octagonal building is 80 feet across (from face to face); the roof is to be a hip roof, and *one-third pitch;* * find the length of the hip rafters.

49. A cylindrical cistern is to have an inside diameter of 20 feet and a depth of 10 feet, and is to be made of concrete 1 foot thick on the bottom; the wall is to be 2 feet thick at the bottom and 1 foot thick at the top, the slope being uniform. Each cubic foot costs 30 cents in place. What will be the cost of the cistern?

50. The chimney, 160 feet high, at a power house, is a frustum of a square pyramid 16 feet square at the base, and 6 feet square at the top; the flue is 3 feet square throughout the entire length. Estimating 19 bricks to the cubic foot, how many will be required to put up the chimney?

51. Before any spoliation, the great pyramid in Egypt had for its base a square each side of which was 763.8 feet and its altitude was 486.26 feet; what was its volume in cubic yards?

52. A cask 3 feet high has a bung diameter of 3.18 feet, and a head diameter of 2.55 feet. A gallon is 231 cubic inches; find the number of gallons.

53. It is proposed to put a dam across a cañon for the purpose of making a reservoir.

The slope of the hill on one side is 1 on 3; on the other 1 on 1. The dirt on the side of the dam away from the water will lie at a slope of 1 on 1, while on the side toward the water the slope will be 1 on 4.

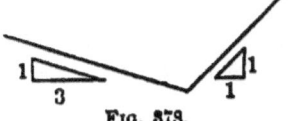

Fig. 378.

The dam is to be 90 feet high with a roadway 12 feet wide on top.

---

* NOTE. — *One-third pitch* indicates a slope of roof such that $a = \dfrac{b}{3}$. The $\angle \theta$ will be somewhat greater than 33° 41'.

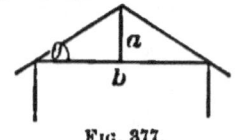

Fig. 377.

It will cost 30 cents per cubic yard to place the dirt in the dam. What will the dam cost if built?

54. A certain reservoir is an inverted frustum of a cone; the upper circumference is 355 metres; the lower circumference is 120 metres; and the length of the slope is 30 metres. Find the capacity of the reservoir in hectolitres.

55. The brick to be used in lining a well are 8″ (inches) × 4″ × 2½″. The well is to be 4′ (feet) in the clear, and 60′ deep. How many brick will be required?

56. A barrel contains 31½ gallons; its bung diameter is 18 inches, and its head diameter is 15 inches. What must be its length?

THEOREMS.

57. If two tetraedrons have a triedral in each equal, the ratio of their volumes equals the ratio of the products of the three edges which determine the angle.

58. The planes which pass through the concurrent edges of a tetraedron and the middle points of the opposite edges, all pass through one line.

59. The lines joining the vertices of a tetraedon with the points of intersection of the medians of the opposite faces, are concurrent; and each is separated into parts which have the ratio of 3 : 1.

NOTE. — This point is the centre of gravity of the tetraedron.

60. The lines joining the middle points of the opposite edges of a tetraedron are concurrent; and the point is the centre of gravity.

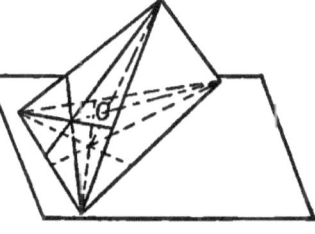

FIG. 879.

61. The altitude of a regular tetraedon equals the sum of the perpendiculars on the four faces, from any point.

62. If a, b, c, and d represent the altitudes of a tetraedron, and a', b', c', and d' the parallel perpendiculars to the faces from any point, then:

$$\frac{a'}{a} + \frac{b'}{b} + \frac{c'}{c} + \frac{d'}{d} = 1.$$

63. A plane angle may be projected into an equal angle, a smaller angle, or a greater angle.

64. Any plane area orthogonally projected on a plane, will bear to the projected area the same ratio that a straight line in the given plane area perpendicular to the intersection of the two planes, bears to the projection of that line.

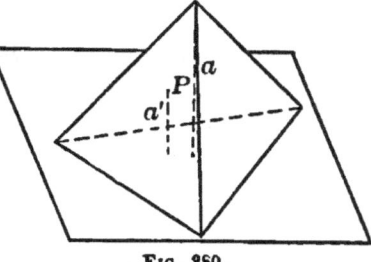

Fig. 380.

NOTE. — The ratio of the projected segment of a line divided by the segment projected is called the cosine of the angle of inclination of the two lines, as the student will see when he begins the study of Trigonometry.

It is customary to say "a projected plane area equals the given area multiplied by the cosine of the angle of inclination."

65. If a tetraedron is cut by a plane parallel to a pair of opposite edges, the section will be a parallelogram, the area of which will be a maximum when the parallelogram is a middle section.

66. The six planes which bisect the six diedrals of a tetraedron all pass through one point.

67. Through any four points, one, and only one, sphere may be passed.

68. If a circle be circumscribed about each face of a tetraedron and perpendiculars be erected at the centres, they will be concurrent.

69. Three spheres intersect each other; the planes of their intersection pass through a common line; and the tangents to the three spheres from any point in this line will be equal to each other.

PROBLEMS.

70. The same number expresses the volume and the surface of a sphere. What is its radius?

71. Show that the surface generated by a triangle revolving about one side equals the length of the two generating sides, multiplied by the path described by the centre of gravity.

72. Show that the volume generated by a triangle revolving about one side, equals the area of the triangle, multiplied by the distance traversed by the centre of gravity.

73. Tillamook light is 138 feet above high water; the lookout on board ship is 60 feet from the water; the tide is 12 feet. What is the difference of the distances at which the light can be seen from the ship at high tide and at low tide?

74. Find the polyedron that would be formed by joining the middle points of the faces of:

(a) A tetraedron,   (c) An octaedron,
(b) A cube,   (d) A dodecaedron,
(e) An icosaedron.

75. Show how to find approximately the diameter of the earth by comparing the angles of elevation of the north pole as observed at two points, A and B, which are on the same meridian and a known distance from each other.

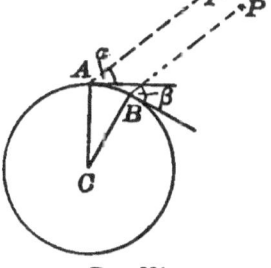

Fig. 381.

NOTE. — The distances of the fixed stars from the earth are so great that lines from any one of them to different points on the earth are practically parallel.

76. Show that if through any point on the surface of a sphere, three chords be drawn at right angles to each other, the sum of the squares of these chords equals the square of the diameter.

77. A sphere 8 inches in diameter has a cylindrical hole 4 inches in diameter bored through it, the axis of the cylinder passing through the centre of the sphere. What is the volume that remains?

78. Having given the radius of the inscribed sphere, find the volumes of:

    (*a*) The circumscribed regular tetraedron,

    (*b*) The circumscribed regular hexaedron,

    (*c*) The circumscribed regular octaedron,

    (*d*) The circumscribed regular dodecaedron,

    (*e*) The circumscribed regular icosaedron.

79. Show geometrically that:

$$(a+b)^3 = a^3 + 3a^2b + 3ab^2 + b^3.$$

80. Show that the area of a zone of one base equals the area of the circle which has the chord of the generating arc for its radius.

81. Show that the volume between the surfaces of two concentric spheres (a spherical shell) equals the frustum of a right circular cone, the lower base of which has for its radius, the radius of the outer sphere; for the radius of the upper base, the radius of the inner sphere; and for its altitude, four times the difference of the radii.

82. Show that when the frustum of a pyramid or cone is being considered:

$$\frac{H}{6}(B + B' + 4M) = \frac{H}{3}(B + B' + \sqrt{BB'}).$$

83. Apply the prismoidal formula to the determination of the volumes of prolate and oblate spheroids.

84. An elliptical reservoir has for its major and minor diameters 200 and 100 feet respectively; the uniform depth is 20 feet. What will be the contents in gallons?

85. A vault has a rectangular floor 100 × 30 feet, and an arched ceiling 15 feet high, which comes down to the floor; the arch is parabolic; the ends are plane. Find the cubic contents.

Fig. 382.

LOCI.

86. Find the locus of the vertex of spherical triangle, the vertical angle being equal to the sum of the other two.

87. Find the locus of points in space, such that the ratio of their distances from two fixed points is constant.

NOTE. — This is the problem of the lights not confined to a plane.

88. Find the locus of a point in space which moves so that the sum of its distances from two fixed points always remains the same.

89. Find the locus of the centre of a sphere which is tangent to a given plane, and the surface of which passes through a fixed point.

90. Find the locus of a point which moves so that the ratio of its distances from a fixed point and from a fixed plane is constant and less than 1.

91. Find the locus of points, the ratio of the distances of which from a given cylinder and from a given circumference, is 1; the centre of the circle being in the axis of the cylinder, the radius greater than that of the cylinder, and the plane perpendicular to the axis of the cylinder.

THE END.

www.ingramcontent.com/pod-product-compliance
Lightning Source LLC
Chambersburg PA
CBHW022102230426
43672CB00008B/1254